Machine Learning Algorithms and Applications in Engineering

Machine Learning (ML) is a subfield of artificial intelligence that uses soft computing and algorithms to enable computers to learn on their own and identify patterns in observed data, build models that explain the world, and predict things without having explicit pre-programmed rules and models. This book discusses various applications of ML in engineering fields and the use of ML algorithms in solving challenging engineering problems ranging from biomedical, transport, supply chain and logistics, to manufacturing and industrial. Through numerous case studies, it will assist researchers and practitioners in selecting the correct options and strategies for managing organizational tasks.

Features:

- Includes the latest research contributions on ML perspectives and applications.
- Follows an algorithmic approach for data analysis in ML and includes real-time engineering case studies.
- Addresses the emerging issues in computing such as deep learning, Internet of Things, and data analytics.
- Focuses on ML techniques that are unsupervised and semi-supervised for unseen and seen data sets on commercial value-added research applications.
- Discusses how the advancements on data science and computing domain open possibilities for cross-disciplinary connections.

This book will benefit professionals, academic researchers, and graduate students in computer science and engineering, biomedical, mechanical, electrical, transport, manufacturing, and industrial engineering.

Machine Learning Algorithms and Applications in Engineering

Edited by
Prasenjit Chatterjee, Morteza Yazdani,
Francisco Fernández-Navarro, and
Javier Pérez-Rodríguez

CRC Press
Taylor & Francis Group
Boca Raton London New York

CRC Press is an imprint of the
Taylor & Francis Group, an **informa** business

First edition published 2023
by CRC Press
6000 Broken Sound Parkway NW, Suite 300, Boca Raton, FL 33487-2742

and by CRC Press
4 Park Square, Milton Park, Abingdon, Oxon, OX14 4RN

CRC Press is an imprint of Taylor & Francis Group, LLC

© 2023 Taylor & Francis Group, LLC

ISBN: 9780367569129 (hbk)
ISBN: 9780367612559 (pbk)
ISBN: 9781003104858 (ebk)

DOI: 10.1201/9781003104858

Typeset in Palatino
by Newgen Publishing UK

Contents

Preface

Some of the decisions taken in engineering, from bioinformatics to risk analysis, are supported by information obtained from data. In recent decades, both academic and private institutions have sought to agglomerate that information in data centers, aiming to extract representative knowledge out of it. In particular, the improvement in information-storage resources has reached a point where almost every institution may store and preserve a considerable amount of data related to its daily services. As expected, raw data stored in those data centers is useless and requires analysis through advanced algorithms. During the twenty-first century, these algorithms have had a remarkable development, outperforming classical statistical methods. Machine learning (ML) and big data technologies have become the solution to massive information processing needs that exceed the capabilities of conventional computing. They are characterized by three features: volume (a large amount of data is produced), velocity (the speed at which these data are obtained and accessed), and variety (data diversity as they come from different sources; e.g., structured, unstructured, multimedia, textual). In this book, we will present several novel applications of ML approaches to engineering problems extremely relevant in practice, ranging from health care to risk analysis.

Organization of the Book

This book is organized into 18 chapters. A brief description of each chapter is presented below.

Chapter 1 reports successful ML models in smart health care through clinical applications such as medical diagnostics and precision or monitoring health, their potential, and limitations. With the increasing power of supercomputers, ML developments would have the proper environment for managing health such as detecting diseases precisely and early, improving diagnosis, and better therapies.

Chapter 2 proposes predictive ML models for assisting exact foreseeing models in recognizing and mapping flood hazard territories. The calculations obtained through ML techniques lead to hazard exhaustion, arrangement suggestion, alleviating loss of helpful lives, and decreasing flood-related property harm.

The main purpose of Chapter 3 is to identify the flaws in current air quality and to recommend measures or policies to improve it so that it is safe to consume. Because each country employs a different scale for evaluation, the data provided by the Central Pollution Control Board of India are utilised to analyse the level of air in the zone under investigation.

Chapter 4 implements an expert system model for solving the problem associated with the precise prediction of the dynamic trajectory of an autonomous vehicle. This was accomplished by deriving a new equation for determining the lateral tire forces and adjusting some of the vehicle parameters under road test conductions. A universal approach to performing the reverse engineering of electric power steering (EPS) for external control is also presented in this chapter.

Chapter 5 proposes a speech recognition model that compiles contemporary expertise on the detection of gestures and their associated mannerisms while proposing modifications for inclusivity of a wider audience, with a prominent emphasis on people with communication disabilities. Furthermore, the work done in this chapter may culminate in the development of real-world applications such as determining the attentivity of students in real time.

Chapter 6 covers the fundamental phenomenon of brain–computer interface as an emerging technology that may help to become possible to communicate dreamers into their real dream through modern technologies. This chapter contains only the essential concepts and theories in integration for developing a deep learning-based model for mapping dream contents. The first part of the chapter narrates dream theories and related cortical regions, whereas the second one concentrates briefly on pillar technology brain–computer interface.

Chapter 7 proposes the use of ML algorithms to classify breast cancer more efficiently through the exploitation of existing data on cancer. This chapter examines the cancer patient data obtained from images, how the data is preprocessed, how different ML models are trained, and chooses the best-trained model for the classification of breast cancer. The current study helps not only in identifying the best ML model to classify breast cancer but also to formulate a process of diagnosing breast cancer based on the preprocessed data obtained from the raw images of breast cancer.

Chapter 8 focuses on the prediction of protein functions, addressing the problem from a computer science (machine learning-based) perspective. Indeed, it can be seen as an overview on the topic of the prediction of protein function aimed especially at computer scientists and researchers who work on machine learning. It will attempt to formally define the multilabel problem of protein function prediction, to explain its magnitude and importance, and why it remains as an open and challenging problem nowadays.

In Chapter 9, the profile injection attack methods, collaborative filtering in recommender systems, and shilling attack detection schemes are discussed from an ML perspective. The shilling attack detection technique works on two parameters, namely, rating parameter and rating and time interval parameter. Based on the working parameters, the shilling attack detection techniques have been classified. They have also been classified based on their output. A brief discussion on every scheme has also been presented.

Chapter 10 highlights some of the primary time series data mining (TSDM) tasks, among which time series preprocessing, segmentation, or prediction are widespread in the literature. ML techniques along with other time series approaches are detailed throughout a set of real-world applications, such as the wave height time series reconstruction, the detection, and prediction of tipping points in paleoclimatology time series, the forecasting of low-visibility events produced by the existence of fog or convective situations, among others.

Chapter 11 uses ML algorithm for predicting quality of a product in terms of minimum shrinkage ratio produced by selective laser sintering (SLS) using some key parameters including surrounding working temperature, laser scanning speed, layer thickness, scanning mode, hatch distance, laser power, and interval time.

Chapter 12 proposes class batch pid partitioning (CBPP) algorithm to provide an optimal balance between utility and privacy for data publishing of 1:M micro data with multiple sensitive attributes.

Chapter 13 focuses on assessing the performance data obtained from real networks during the COVID-19 pandemic using scripts and existing applications for monitoring apps and networks in real time.

Chapter 14 highlights the integration of AI/ML for 5G network considering different constraints.

In Chapter 15, ML methods are used to forecast power prices using an open-source dataset. For better prediction and visualization of electricity prices than the already anticipated values in the dataset, a web tool called Jupyter notebook is employed.

In Chapter 16, ML is used to forecast the qualities of various materials as well as their deterioration rates. This model also generates the composition of materials, as well as their mechanical and biological properties, saving time and reducing failures during experimental examination.

The objective of Chapter 17 is to predict the various outcomes of Myocardial Infarction (commonly known as a heart attack) using Neural Decision Forest.

The goal of Chapter 18 is to show a current implementation of SimCLR, a basic framework for contrastive learning of visual representations, which is a contrastive self-supervised learning method. This research looks into how to learn these visual representations by examining the framework's most critical sections on a regular basis.

The Editors

Prasenjit Chatterjee is currently the Dean (Research and Consultancy) at MCKV Institute of Engineering, West Bengal, India. He has published over 120 research papers in various international journals and peer-reviewed conferences. He has authored and edited more than 22 books on intelligent decision-making, supply chain management, optimization techniques, and risk and sustainability modeling. He has received numerous awards including Best Track Paper Award, Outstanding Reviewer Award, Best Paper Award, Outstanding Researcher Award, and university gold medal. Chatterjee is the Editor-in-Chief of the *Journal of Decision Analytics and Intelligent Computing*. He has also been the guest editor of several special issues in different SCIE / Scopus / ESCI (Clarivate Analytics) indexed journals. He is also the Lead Series Editor of "Smart and Intelligent Computing in Engineering" (Chapman and Hall / CRC Press), Founder and Lead Series Editor of "Concise Introductions to AI and Data Science" (Scrivener–Wiley); *AAP Research Notes on Optimization and Decision Making Theories; Frontiers of Mechanical and Industrial Engineering* (Apple Academic Press, co-published with CRC Press, Taylor and Francis Group), and "River Publishers Series in Industrial Manufacturing and Systems Engineering." Chatterjee is one of the developers of two multiple-criteria decision-making methods called Measurement of Alternatives and Ranking according to Compromise Solution (MARCOS) and Ranking of Alternatives through Functional mapping of criterion sub-intervals into a Single Interval (RAFSI).

Morteza Yazdani currently works at Universidad Autónoma de Madrid, Spain, as an assistant professor. He also worked at ESIC University and Universidad Loyola Andalucía, Seville, for several years. Yazdani's major researches focus on management and operations, specifically on decision-making theories, supply chain management, and sustainable development. Yazdani is a member of the editorial board of *International Journal of Productivity and Performance Management*. He teaches operations management and statistics, among other subjects. Yazdani has published in journals like *International Journal of Production Research, Expert Systems with Applications, Computers and Industrial Engineering*, and *Annals of Operations Research*.

Javier Pérez Rodríguez is a computer science engineer (University of Córdoba), with a master's degree in soft computing and intelligent systems (University of Granada) and doctorate in information and communication technologies from the University of Granada. His area of research and teaching are in two areas: computer science and artificial intelligence and bioinformatics. Pérez has published 13 articles in the best journals in both areas, all of them with a high impact index (Q1). His teaching experience has been mainly in different and diverse undergraduate subjects within the official teaching load assigned to the Department of Quantitative Methods of Loyola Andalucía University.

Francisco Fernández-Navarro is an associate professor at the Universidad Loyola Andalucía and a member of IEEE Computer Society. Formerly, he was a research fellow, Innovation Dynamics and Computational Economics, Advanced Concepts Team at the European Space Agency in the Netherlands. He got his PhD in artificial intelligence and software engineering from the University of Malaga. He is teaching forecasting and machine learning, mathematics, and statistics, among others. Fernández Navarro has published in top journals like *Expert Systems with Applications, Applied Soft Computing,* and *Neural Computing and Applications.* He has supervised several PhD theses as well.

1

Machine Learning for Smart Health Care

Rehab A. Rayan
Department of Epidemiology, High Institute of Public Health,
Alexandria University, Egypt

rayanr@alexu.edu.eg

CONTENTS

1.1   Introduction

Machine learning (ML) incorporates advanced algorithms working on disparate big data to reveal valuable trends that would be challenging to be figured out by even skilled experts. Nowadays, ML applications spawn several disciplines such as gaming (Silver et al. 2018), suggesting products (Batmaz et al. 2019), and self-driven cars (Bojarski et al. 2016) while in medicine, the examples are the human genome project (Venter et al. 2001) and the cancer omics (genomics and proteomics) (Zhang et al. 2019; Ellis et al. 2013). Gathering and exploring health-related big data has the potential to shift medicine into a data-centered and outcomes-focused field with advances in detecting, diagnosing, and managing diseases. Molecular and phenotypic datasets were developed, which covered genetic examination for individualized cancer therapy, high-resolution three-dimensional anatomical body parts' images, histological examination of biopsies, and smartwatches with biosensors for monitoring heart rates and alerting about abnormalities

DOI: 10.1201/9781003104858-1

(Shilo et al. 2020). Such big data supply the raw material for a future of early, precise diagnosis, customized therapies, and continuous monitoring of health.

ML could promote health care by releasing the potential of health big data. Prior applications of ML in diagnosis and care were promising in detecting breast cancer from X-rays (McKinney et al. 2020), finding novel antibiotics (Stokes et al. 2020), expecting gestational diabetes early from digital medical records (Artzi et al. 2020), and determining patients who share a molecular signature of therapeutic responses (Zitnik et al. 2019). ML digital pattern recognition could tackle complicated health big data in cases where a manual exploration is infeasible and ineffective. Several diseases include complicated modifications, which are experienced robustly and differently among patients and need diligent exploration and cautious evaluation of disparate big data to ascertain unique patterns so as to diagnose and manage, hence, helping health care providers and researchers to find and describe valuable insights from such big data (Rajkomar et al. 2019). The functionality of a new algorithm could be strictly evaluated with previously proven associations between either quantitative biomarkers or qualitative data (which vary based on demographics and ecological exposures) and patient health outcomes.

More research models have been developed for gathering and assembling big data that relate attributes to a health condition, which could be applied in training and testing ML techniques. Cancer models could accumulate molecular profiles from testing frameworks or patients' samples with data on diagnosis, treatment, and prognosis; for instance, the Cancer Dependency Map has gathered data on genomic stability, multimodal molecular profiles, and therapeutic responses from thousands of cancer cell lines. By applying innovative algorithms, these models could bring about a shift in knowledge about illnesses and advance anticipating health outcomes (Tsherniak et al. 2017).

ML is derived from artificial intelligence (AI), which involves techniques to facilitate machines showing learning and reasoning similar to humans. ML stresses on building algorithms to learn from data. Major ML classes involve supervised learning where datasets are linked to a certain outcome; categorical values use classification techniques such as "healthy" or "diseased" while continuous ones are applied in regression models such as responding to treatment levels; semi-supervised or unsupervised techniques to classify data into specific sets, which could be manually tagged and linked to an outcome; ensemble learning, in which findings from many digital frameworks are integrated to give a final recommendation enabling more precise estimations via facilitating frameworks that enhance scaling-up to novel data; deep learning (DL) that applies artificial neural networks, inspired by similar networks in the human brain, to identify trends and relations in the data where it is valuable operating on unstructured data like text, speech, or images; and Bayesian learning where previous information is encrypted into the procedure of learning, particularly while working on poor data (Goecks et al. 2020).

Several health datasets are big, having many dimensions. Reducing the dimensions of datasets could enhance the functionality of ML techniques by choosing a subgroup of proper data variables or merging variables into a lower number, obtaining data variance, visualizing data, or predicting models. In ML, techniques for federated learning could gradually learn data that are spread over many locations and cannot be merged into one dataset (Yang et al. 2019). Federated learning is crucial in several medical applications working with protected or sensitive health data. Lately, accessing structured health data and technical advances in data analytics tools contributed to the growth of these techniques (Camacho et al. 2018). The ML techniques are selected according to the accessed data along with the applications they are intended for.

With the growing powerful ML supercomputers and the infrastructure for gathering and analyzing data precisely, ML techniques would play key roles in managing health such as detecting diseases precisely and early, improving diagnosis, and better therapies. The quality of the learned data could be advanced greatly through adopting standardized and large-scale electronic health records (EHRs) developed particularly for ML. Home data collection via smartphones, home-assisting devices such as Google Home or Amazon Echo, and other digital devices would contribute to more advanced ML in health. Figure 1.1 depicts how sophisticated algorithms are applied on the health big data for smart outcomes-based individualized health care. Whereas in a house, ML could early detect disease, monitor compliance, and response to therapy, in a hospital, ML could enable health care providers to diagnose and personalize care for a freely moving patient between the different settings.

This chapter discusses how ML could promote health care via better medical diagnostics, precision health, and monitoring health, highlighting opportunities and successful early applications, and concluding with challenges that hinder achieving the full potential of ML in health.

1.2   Major Applications of ML in Health Care

1.2.1   Medical Diagnostics

Today, medical examinations yield more data than ever thanks advances in medical care. High-fidelity imaging scans generate huge two-, three-, or four-dimensional images for tissues and organs, and molecular examinations could analyze hundreds to thousands of genes or proteins. ML could digitally analyze the diagnostic characteristics of such data to predict disease or

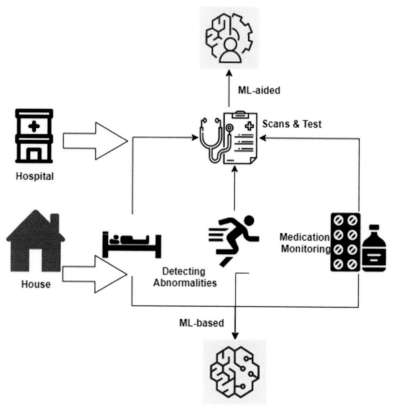

FIGURE 1.1
ML applications in health.

response to therapy. Deep learning (DL) could analyze and interpret medical images where many late pieces of research indicated that programs applying ML such as computer-aided detection (CAD) could interpret radiological scans comparable to radiologists. For instance, a DL-based CAD program could precisely detect diabetic retinopathy (Gulshan et al. 2016) and determine all grades in situ or invasive breast cancer comparable to medical experts (McKinney et al. 2020). Therefore, with the help of big data, DL-based techniques could function similarly to professionals over a spectrum of diagnosing jobs in medical imaging with high precision (Liu et al. 2019).

Molecular assays could determine abnormalities in genes and measure stages of gene expression and abundant proteins in different samples such as tissues, saliva, or blood where ML could identify complicated biomarkers linked to different diseases, informing patients' outcomes, and determining proper managing planes. In cancer biology, for instance, applying positioning nucleosome (Heitzer et al. 2019) and methylating DNA (Kang et al. 2017) in the blood to estimate the source tissue of a tumor, measuring the stage of induction in the cell pathway from biopsies or other samples (Way and Greene 2018), applying magnetic resonance imaging to estimate genetic characteristics of brain tumors (P. Chang et al. 2018), and integrating imaging with omics to predict outcomes in cancer patients (Chaudhary et al. 2018; Mobadersany et al. 2018). ML also could identify patients who are deprived of sleep via exploring blood mRNA, indicating the adverse effect of inadequate sleep on health (Laing et al. 2019). With the help of coordinated data of different sources and biomarkers, the ML models are promising to work more precisely than current techniques that are usually confined to a low set of biomarkers reflecting only limited insights of the complicated diseases.

Cooperative human–machine diagnostic techniques, as shown in Figure 1.1, are promising for taking the advantages of both humans and machines where the health care provider could arrive at a diagnosis via combining all the accessed data including those generated through ML platforms (Ahuja 2019). Hence, ML would computerize regular diagnosis, labeling troublesome cases in need for further human intervention, and offering more valuable data to reach a diagnosis (Ardila et al. 2019). Therefore, integrating knowledge from both health care experts and innovative algorithms would advance diagnoses however. There is a need to evaluate the biologic utility of these functionalities. Hence, prior to broad installation and implementation, it is vital to control the transparency of ML applications via optimal goals, qualities, measurable functionalities, and constraints of certain algorithms and their verifying processes (Cai et al. 2019). Thus, assisting health care providers in applying ML programs to reach correct conclusions enhances decision-making. ML-based programs could grow confidence in the health care system, enabling more knowledge about the implicit biologic pathways in illnesses (Ching et al. 2018).

ML, coupled with more advances in techniques for clinical investigations, requires considering the balance between detection rates for diseases, patient outcomes, and distinct elements influencing health and quality of life. Applying ML techniques would elevate the rate of detecting diseases, and differentiating severe from minor diseases would be necessary to determine subgroups of diseases, choose highly effective therapies, and eliminate overtreatment. For ML to promote health care, thoughtfully formulating clinical objectives and the related testing and validation measures are required.

1.2.2 Precision Health

In precision health, a promising ML application, health care is customized as per the patient's disease profile. In precision oncology, an initial model for ML in precision health, cancer is managed according to the molecular features of the tumor. Lately, molecular biomarkers for a subject like the levels of gene expressions or physical transformations usually inform the choice of therapies, yet variations in epigenomic, genomic, and anatomic disease characteristics lead to largely different individual responses (Brown et al. 2019). Besides, the hundreds of likely medications and therapeutic regimens may not be suitable for all conditions (Kurnit et al. 2018).

Multiscale predictive frameworks are promising ML techniques in dealing with subjective variability. For instance, single-purpose frameworks could predict the functioning outcomes of biologic alterations, like the effect of gene mutations on splicing and genetic expression (Xiong et al. 2015) and transcription element-binding (K. M. Chen et al. 2019). ML techniques could also forecast therapeutic response in cancer cell lines (Y. Chang et al. 2018), translate forecasting to patient tumors (Chiu et al. 2019), and predict response to treatment according to clinical data on outcomes (Huang et al. 2018). Upcoming innovations in modeling for precision medicine are promising for the work on several dimensions for several needs. Multiscale modeling would use biological big data to analyze an organism growing through unique aspects of time and space. Now there are digital frameworks of human–virus interplay (Lasso et al. 2019), cell-to-cell interactions between tumor cells and immune cells, or the entire cells (Metzcar et al. 2019). Hence, digital frameworks of organs or even the whole organism, known as digital twins (Björnsson et al. 2019), hold the potential for future where they could forecast the potency of various medication mixtures, which have never been tried in combination before and model the effect of illnesses on various organs.

However, such multifactorial frameworks would aid in precise predictions to be directly applied in therapy, and they might reach a mid-level where ML techniques create a graded roster of recommended treatments guiding skillful and qualified health care providers in decision-making. For example, patient-centered laboratory frameworks could test predictions from digital models and suggest the best-functioning ones for therapy. Such a hybrid technique is promising where ML frameworks could significantly limit the likely therapeutic combinations to be regarded and determine those to be disregarded. Adding an experimental verifying stage could lend more proof to the potential effectiveness of the predicted treatments.

Applying ML in digitally searching and mining knowledge in patients' datasets and publications would advance precision health (Rajkomar et al. 2019). Patient datasets, often found in the EHRs, are a wealthy resource for data on diagnosis, therapies, and responses for vast sets of patients. Previous attempts to apply the algorithms of natural language processing for mining published literature (Dong et al. 2018), EHRs (Shickel et al. 2018), and medical notes (Kreimeyer et al. 2017) yielded interesting information about relations between biologic mechanisms and biomarker-based treatment or using structured EHRs data to forecast the onset of disease (Artzi et al. 2020). ML could leverage such knowledge for applications in precision health via superior techniques, which could handle EHRs and published works of literature's unstructured data and metadata. However, the existing data is incomplete or inaccurate, making the generation of meaningful relations difficult.

1.2.3 Monitoring Health

Treating complex diseases would be shifted from curative toward managing approaches. Such a wide approach for managing health would preserve health across several diseases and the natural aging process. Managing health would need continuous monitoring of all health aspects for likely diseases and personalizing therapies according to patients' responses and hence the key role of ML. Figure 1.2 shows integrating data and ML for inclusive, ongoing, and precision health monitoring where data collected at both houses or clinics are combined through predictive frameworks. Inclusive frameworks are promising for better performance since they integrate more personal data and are adaptable in any setting.

Beyond the clinical environment, wearables and intelligent home electronics could be used to manage health by gathering vast quantities of high-quality health-related data to be used by ML techniques for recommending timely interventions, changing lifestyle, or referring to a health care provider for consultations. Today, wearables have in-built biosensors to monitor movement, pulsation, rates of respiration, and levels of oxygen, body temperature, and blood pressure, among other indicators. Test models showed

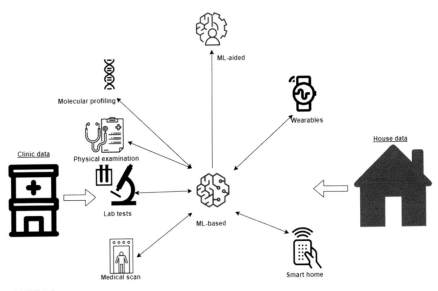

FIGURE 1.2
Inclusive ML framework.

that data from wearables could be used for managing diabetes (Chang et al. 2016), detecting atrial fibrillation (Bumgarner et al. 2018) or early diagnosis of Parkinson's disease (Lonini et al. 2018), tracking cholesterol levels in the blood (Fu and Guo 2018), complying with therapeutic regimens (Car et al. 2017), and to prompt an alert on a cardiac arrest (Sahoo, Thakkar, and Lee 2017). Speech-based home aids could identify agonal breathing, an audible early sign of a heart attack (Chan et al. 2019). Soon, ML applications would be promising in detecting more biomarkers from audio and wearable sensors' data, maybe via integrating data among various platforms where DL and classical supervised learning might design models from such data.

Applying ML to the gathered data via smartphones is also promising for diagnostics. DL techniques could analyze smartphone-captured photos to detect various forms of dermal tumors (Esteva et al. 2017) and to identify diabetes retinopathy (Micheletti et al. 2016). Lately, researchers have found that smartphone-gathered sensory data such as voice, response time, and accelerometer data could be processed applying ML to monitor symptoms and progress of Parkinson's disease (Ginis et al. 2016). These tested models showed that ML-based wearables, home devices, and smartphones could collect valuable data involving biometric measures, images, diet consumption, and ecological data (Vermeulen et al. 2020). Linking such data with diagnoses, ML could recognize patterns within the data and point to a certain diagnosis.

Managing health implies continuous monitoring of the subject's body functions and behavior via wearables and home devices coupled with results from regular blood examinations. Using baseline activities and functions, individualized frameworks could be generated by tailoring population-wide frameworks to gathered personal data, enabling accumulating individual baselines and detecting deviations, which might reflect changes in health. Applying individualized frameworks, ML techniques could monitor all abnormalities in a person and alert them if a consultation with health care providers should be done. Besides, tracking subjects who are searching online for their symptoms, for example, self-reported health problems like losing weight, bronchitis, coughing, chest pain, among others, along with machine-trained multipersonal patterns, could be used to detect pancreas and lung tumors early (White and Horvitz 2017; Paparrizos, White, and Horvitz 2016), hence notifying a health care professional or a patient suggesting seeking medical care if a further severe condition might convey the apparently simple searched symptoms. However, several privacy-related problems are expected.

Using ML, health care providers would be able to interpret the highly precise molecular testing and imaging to recognize significant biomarkers and reach a final diagnosis. Digital search results and multidimensional modeling for dissimilar patients would enable diagnosing diseases requiring therapies and provide informed therapeutic options. Following diagnosis and therapy, managing health starts again with continuous monitoring of personal health.

Yet, an ML framework should monitor personal therapeutic responses, observe any side effects, and monitor general health or unusual deviations from normal. ML would assist in implementing the first individualized framework involving data on diagnosis and treatment, generating an estimated direction for therapy regarded as the new normal.

ML could manage personal health over a lifespan via integrating and modeling data digitally and smartly. Managing health includes creating individualized frameworks for monitoring and precisely discovering abnormalities, supporting health care providers in digital diagnosis and treatment using patients' data, and upgrading personal frameworks with recent diagnoses and therapies. Thus, it is data-intensive process that needs digital pattern for recognizing complex datasets. Managing health would be an ongoing learning process since frameworks would be upgraded with fresh data. Therefore, either creating novel forecasting frameworks or upgrading available ones along with further effort are required for identifying the potentials and barriers of such techniques with various applications.

1.3   Opportunities and Limitations

ML could contribute to transformative approaches in diagnosis and therapy; however, it is required to grow well-structured and high-quality datasets for enhancing the predictive power of ML techniques and limiting the amount of the required data for training and the complexity of the output representations. Speeding up in ML techniques for image recognition happened during the introduction of ImageNet, a group of tagged and ontologically attached images, and likewise efforts in biomedicine are required (Deng et al. 2009).

High-quality datasets for using ML in diagnosis and therapy needs tackling financial, technical, and legal obstacles, which usually lead to scattered and unstandardized data. Federated learning could technically solve this issue via integrating scattered data across systems since actual moving of data is not required and personal privacy could be safeguarded. Home devices and wearables could be used to gather precise data, and ML could process such data to extract precise medical and analytic information from unstructured origins such as published literature and EHRs. Developing regulations for securely managing and analyzing private health information (PHI), and legal and social standards, which determine adherence with them should be created. Both biomedical subjects and entities should be rewarded to participate in standardizing and sharing data. Likewise, drug and insurance companies supporting biomedical research should invest in the infrastructure along with collecting and analyzing data toward better quality datasets.

Rewarding techniques for sharing data, which boost variable datasets for learning, are also required, such as local and global standards for sharing data, which acquire data from both large hospitals and small clinics. ML applications, which enhance patients' therapeutic responses in large hospitals, might not suit small clinics because of variability in patients and general care. Yet, the ideal target of gathering health-related data for ML requires acquiring data from proper representative patients' populations for building precise ML frameworks that can be generalized to the diverse public. Hence, focused attempts considering factors like patient condition before therapy, therapeutic regimes, age, sex, race, ethnicity, and ecological risks are needed (Goecks et al. 2020).

A thorough analysis of ML techniques applied in medicine is required, particularly with ongoing learning. The functionality of an ML system is best tested through the precision of longitudinal predictions. An iterative ML technique might involve training with retrospective data, deploying algorithms, and evaluating precision of the acquired predictions. Deployment-gathered data, and further retrospective big data, could retrain and enhance the algorithm, followed by a cycle of testing for deployment. Assessing ongoing learning systems, like those used in monitoring health, which adapt to changes in behaviors or health conditions, would possibly need strengthening such cycle and applying deployment-gathered data in identifying both barriers and flaws. Besides, measuring confidence intervals is vital, as some ML applications would be further tolerating uncertain predictions than others, hence confidence intervals could inform making decisions.

Iteratively training and deploying applications of ML holds regulatory limitations since the majority of diagnostic and therapeutic applications presume fixed data models. Upgrading models with novel data or adapting them for new diagnoses and therapies would require continuous assessment for the reliability and precision of the predictions. Thus, there is a need for real or simulated datasets, which are prospective and multidimensional and also costly for robustly assessing the applications of ML in medicine.

1.4 Conclusions

ML incorporates advanced algorithms working on disparate big data to reveal valuable trends that would be challenging to be figured out by even skilled experts. ML could promote health care by releasing the potential of health big data. This chapter discussed how ML could promote healthcare via better medical diagnostics and precision or monitoring health, highlighting opportunities and successful early applications, and ending with challenges that hinder achieving the full potential of ML in health. With the growing powerful ML supercomputers and the infrastructure for gathering and analyzing data precisely, ML techniques would play key roles in managing health such as detecting diseases precisely and early, improving diagnosis, and better therapies. The quality of the learned data could be advanced greatly by adopting standardized and large-scale electronic health records (EHRs) developed particularly for ML. Despite considerable limitations, this problem could be addressed. More research models have been developed for gathering and assembling big data that relate attributes to a health condition, which could be applied in training and testing ML techniques. By applying innovative algorithms, these models could bring about a shift in knowledge about illnesses and advance anticipating health outcomes. More efforts are needed for a promising future of rigorous, outcomes-centered health care with detecting, diagnosing, and therapeutic protocols adjusted through ML to personal and ecological diversities and allowing inclusive management of health. ML would assist in the implementation of the first individualized framework involving data on diagnosis and treatment, generating an estimated direction for therapy regarded as the new normal.

References

Ahuja, Abhimanyu S. 2019. "The Impact of Artificial Intelligence in Medicine on the Future Role of the Physician." *PeerJ* 7 (October). doi:10.7717/peerj.7702

Ardila, Diego, Atilla P. Kiraly, Sujeeth Bharadwaj, Bokyung Choi, Joshua J. Reicher, Lily Peng, Daniel Tse, et al. 2019. "End-to-End Lung Cancer Screening with Three-Dimensional Deep Learning on Low-Dose Chest Computed Tomography." *Nature Medicine* 25(6): 954–61. doi:10.1038/s41591-019-0447-x

Artzi, Nitzan Shalom, Smadar Shilo, Eran Hadar, Hagai Rossman, Shiri Barbash-Hazan, Avi Ben-Haroush, Ran D. Balicer, Becca Feldman, Arnon Wiznitzer, and Eran Segal. 2020. "Prediction of Gestational Diabetes Based on Nationwide Electronic Health Records." *Nature Medicine* 26 (1): 71–76. doi:10.1038/s41591-019-0724-8

Batmaz, Zeynep, Ali Yurekli, Alper Bilge, and Cihan Kaleli. 2019. "A Review on Deep Learning for Recommender Systems: Challenges and Remedies." *Artificial Intelligence Review* 52(1): 1–37. doi:10.1007/s10462-018-9654-y

Björnsson, Bergthor, Carl Borrebaeck, Nils Elander, Thomas Gasslander, Danuta R. Gawel, Mika Gustafsson, Rebecka Jörnsten, et al. 2019. "Digital Twins to Personalize Medicine." *Genome Medicine* 12(1): 4. doi:10.1186/s13073-019-0701-3

Bojarski, Mariusz, Davide Del Testa, Daniel Dworakowski, Bernhard Firner, Beat Flepp, Prasoon Goyal, Lawrence D. Jackel, et al. 2016. "End to End Learning for Self-Driving Cars." *ArXiv:1604.07316* [Cs], April. http://arxiv.org/abs/1604.07316.

Brown, Benjamin P., Yun-Kai Zhang, David Westover, Yingjun Yan, Huan Qiao, Vincent Huang, Zhenfang Du, et al. 2019. "On-Target Resistance to the Mutant-Selective EGFR Inhibitor Osimertinib Can Develop in an Allele Specific Manner Dependent on the Original EGFR Activating Mutation." *Clinical Cancer Research*,

January. American Association for Cancer Research. doi:10.1158/1078-0432. CCR-18-3829

Bumgarner, Joseph M., Cameron T. Lambert, Ayman A. Hussein, Daniel J. Cantillon, Bryan Baranowski, Kathy Wolski, Bruce D. Lindsay, Oussama M. Wazni, and Khaldoun G. Tarakji. 2018. "Smartwatch Algorithm for Automated Detection of Atrial Fibrillation." *Journal of the American College of Cardiology* 71 (21): 2381–2388. doi:10.1016/j.jacc.2018.03.003

Cai, Carrie J., Samantha Winter, David Steiner, Lauren Wilcox, and Michael Terry. 2019. "'Hello AI': Uncovering the Onboarding Needs of Medical Practitioners for Human-AI Collaborative Decision-Making." *Proceedings of the ACM on Human–Computer Interaction* 3 (CSCW): 104:1–104:24. doi:10.1145/3359206

Camacho, Diogo M., Katherine M. Collins, Rani K. Powers, James C. Costello, and James J. Collins. 2018. "Next-Generation Machine Learning for Biological Networks." *Cell* 173(7): 1581–1592. doi:10.1016/j.cell.2018.05.015

Car, Josip, Woan Shin Tan, Zhilian Huang, Peter Sloot, and Bryony Dean Franklin. 2017. "EHealth in the Future of Medications Management: Personalisation, Monitoring and Adherence." *BMC Medicine* 15 (April). doi:10.1186/s12916-017-0838-0

Chan, Justin, Thomas Rea, Shyamnath Gollakota, and Jacob E. Sunshine. 2019. "Contactless Cardiac Arrest Detection Using Smart Devices." *NPJ Digital Medicine* 2: 52. doi:10.1038/s41746-019-0128-7

Chang, P., J. Grinband, B. D. Weinberg, M. Bardis, M. Khy, G. Cadena, M.-Y. Su, et al. 2018. "Deep-Learning Convolutional Neural Networks Accurately Classify Genetic Mutations in Gliomas." *AJNR. American Journal of Neuroradiology* 39 (7): 1201–1207. doi:10.3174/ajnr.A5667

Chang, Shih-Hao, Rui-Dong Chiang, Shih-Jung Wu, and Wei-Ting Chang. 2016. "A Context-Aware, Interactive M-Health System for Diabetics." *IT Professional* 18 (3): 14–22. doi:10.1109/MITP.2016.48

Chaudhary, Kumardeep, Olivier B. Poirion, Liangqun Lu, and Lana X. Garmire. 2018. "Deep Learning–Based Multi-Omics Integration Robustly Predicts Survival in Liver Cancer." *Clinical Cancer Research* 24(6): 1248–1259. doi:10.1158/1078-0432. CCR-17-0853

Chen, Kathleen M., Evan M. Cofer, Jian Zhou, and Olga G. Troyanskaya. 2019. "Selene: A PyTorch-Based Deep Learning Library for Sequence Data." *Nature Methods* 16(4): 315–318. doi:10.1038/s41592-019-0360-8

Ching, Travers, Daniel S. Himmelstein, Brett K. Beaulieu-Jones, Alexandr A. Kalinin, Brian T. Do, Gregory P. Way, Enrico Ferrero, et al. 2018. "Opportunities and Obstacles for Deep Learning in Biology and Medicine." *Journal of the Royal Society Interface* 15 (141): 20170387. doi:10.1098/rsif.2017.0387

Chiu, Yu-Chiao, Hung-I Harry Chen, Tinghe Zhang, Songyao Zhang, Aparna Gorthi, Li-Ju Wang, Yufei Huang, and Yidong Chen. 2019. "Predicting Drug Response of Tumors from Integrated Genomic Profiles by Deep Neural Networks." *BMC Medical Genomics* 12(1): 18. doi:10.1186/s12920-018-0460-9

Deng, Jia, Wei Dong, Richard Socher, Li-Jia Li, Kai Li, and Li Fei-Fei. 2009. "ImageNet: A Large-Scale Hierarchical Image Database." In *2009 IEEE Conference on Computer Vision and Pattern Recognition*, 248–255. doi:10.1109/CVPR.2009.5206848

Dong, Wei, Xiaoling Wang, Zhi Xia, Xiuqing Zhang, and Huanming Yang. 2018. "A Legacy of the '1% Program'—The 'Chinese Chapter' of the Human Genome Reference Sequence." *Journal of Genetics and Genomics—Yi Chuan Xue Bao* 45(11): 565–568. doi:10.1016/j.jgg.2018.10.003

Ellis, Matthew J., Michael Gillette, Steven A. Carr, Amanda G. Paulovich, Richard D. Smith, Karin K. Rodland, R. Reid Townsend, et al. 2013. "Connecting Genomic Alterations to Cancer Biology with Proteomics: The NCI Clinical Proteomic Tumor Analysis Consortium." *Cancer Discovery* 3 (10): 1108–1112. doi:10.1158/2159-8290.CD-13-0219

Esteva, Andre, Brett Kuprel, Roberto A. Novoa, Justin Ko, Susan M. Swetter, Helen M. Blau, and Sebastian Thrun. 2017. "Dermatologist-Level Classification of Skin Cancer with Deep Neural Networks." *Nature* 542 (7639): 115–118. doi:10.1038/nature21056

Fu, Yusheng and Jinhong Guo. 2018. "Blood Cholesterol Monitoring with Smartphone as Miniaturized Electrochemical Analyzer for Cardiovascular Disease Prevention." *IEEE Transactions on Biomedical Circuits and Systems* 12(4): 784–790. doi:10.1109/TBCAS.2018.2845856

Ginis, Pieter, Alice Nieuwboer, Moran Dorfman, Alberto Ferrari, Eran Gazit, Colleen G. Canning, Laura Rocchi, Lorenzo Chiari, Jeffrey M. Hausdorff, and Anat Mirelman. 2016. "Feasibility and Effects of Home-Based Smartphone-Delivered Automated Feedback Training for Gait in People with Parkinson's Disease: A Pilot Randomized Controlled Trial." *Parkinsonism & Related Disorders* 22 (January): 28–34. doi:10.1016/j.parkreldis.2015.11.004

Goecks, Jeremy, Vahid Jalili, Laura M. Heiser, and Joe W. Gray. 2020. "How Machine Learning Will Transform Biomedicine." *Cell* 181 (1): 92–101. doi:10.1016/j.cell.2020.03.022

Gulshan, Varun, Lily Peng, Marc Coram, Martin C. Stumpe, Derek Wu, Arunachalam Narayanaswamy, Subhashini Venugopalan, et al. 2016. "Development and Validation of a Deep Learning Algorithm for Detection of Diabetic Retinopathy in Retinal Fundus Photographs." *JAMA* 316 (22): 2402–2410. doi:10.1001/jama.2016.17216

Heitzer, Ellen, Imran S. Haque, Charles E. S. Roberts, and Michael R. Speicher. 2019. "Current and Future Perspectives of Liquid Biopsies in Genomics-Driven Oncology." *Nature Reviews Genetics* 20 (2): 71–88. doi:10.1038/s41576-018-0071-5

Huang, Cai, Evan A. Clayton, Lilya V. Matyunina, L. DeEtte McDonald, Benedict B. Benigno, Fredrik Vannberg, and John F. McDonald. 2018. "Machine Learning Predicts Individual Cancer Patient Responses to Therapeutic Drugs with High Accuracy." *Scientific Reports* 8 (1): 16444. doi:10.1038/s41598-018-34753-5

Kang, Shuli, Qingjiao Li, Quan Chen, Yonggang Zhou, Stacy Park, Gina Lee, Brandon Grimes, et al. 2017. "CancerLocator: Non-Invasive Cancer Diagnosis and Tissue-of-Origin Prediction Using Methylation Profiles of Cell-Free DNA." *Genome Biology* 18(1): 53. doi:10.1186/s13059-017-1191-5

Kreimeyer, Kory, Matthew Foster, Abhishek Pandey, Nina Arya, Gwendolyn Halford, Sandra F. Jones, Richard Forshee, Mark Walderhaug, and Taxiarchis Botsis. 2017. "Natural Language Processing Systems for Capturing and Standardizing Unstructured Clinical Information: A Systematic Review." *Journal of Biomedical Informatics* 73: 14–29. doi:10.1016/j.jbi.2017.07.012

Kurnit, Katherine C., Ecaterina E. Ileana Dumbrava, Beate Litzenburger, Yekaterina B. Khotskaya, Amber M. Johnson, Timothy A. Yap, Jordi Rodon, et al. 2018. "Precision Oncology Decision Support: Current Approaches and Strategies for the Future." *Clinical Cancer Research: An Official Journal of the American Association for Cancer Research* 24(12): 2719–2731. doi:10.1158/1078-0432.CCR-17-2494

Laing, Emma E., Carla S. Möller-Levet, Derk-Jan Dijk, and Simon N. Archer. 2019. "Identifying and Validating Blood MRNA Biomarkers for Acute and Chronic Insufficient Sleep in Humans: A Machine Learning Approach." *Sleep* 42 (1). doi:10.1093/sleep/zsy186

Lasso, Gorka, Sandra V. Mayer, Evandro R. Winkelmann, Tim Chu, Oliver Elliot, Juan Angel Patino-Galindo, Kernyu Park, Raul Rabadan, Barry Honig, and Sagi D. Shapira. 2019. "A Structure-Informed Atlas of Human-Virus Interactions." *Cell* 178(6): 1526–1541.e16. doi:10.1016/j.cell.2019.08.005

Liu, Xiaoxuan, Livia Faes, Aditya U. Kale, Siegfried K. Wagner, Dun Jack Fu, Alice Bruynseels, Thushika Mahendiran, et al. 2019. "A Comparison of Deep Learning Performance against Health-Care Professionals in Detecting Diseases from Medical Imaging: A Systematic Review and Meta-Analysis." *The Lancet Digital Health* 1(6): e271–e297. doi:10.1016/S2589-7500(19)30123-2

Lonini, Luca., Dai, Andrew., Shawen, Nicholas. *et al.* Wearable sensors for Parkinson's disease: which data are worth collecting for training symptom detection models. *npj Digital Med* 1, 64 (2018). https://doi.org/10.1038/s41746-018-0071-z

McKinney, Scott Mayer, Marcin Sieniek, Varun Godbole, Jonathan Godwin, Natasha Antropova, Hutan Ashrafian, Trevor Back, et al. 2020. "International Evaluation of an AI System for Breast Cancer Screening." *Nature* 577 (7788): 89–94. doi:10.1038/s41586-019-1799-6

Metzcar, John, Yafei Wang, Randy Heiland, and Paul Macklin. 2019. "A Review of Cell-Based Computational Modeling in Cancer Biology." *JCO Clinical Cancer Informatics* 3: 1–13. doi:10.1200/CCI.18.00069

Micheletti, J. Morgan, Andrew M. Hendrick, Farah N. Khan, David C. Ziemer, and Francisco J. Pasquel. 2016. "Current and Next Generation Portable Screening Devices for Diabetic Retinopathy." *Journal of Diabetes Science and Technology* 10(2): 295–300. doi:10.1177/1932296816629158

Mobadersany, Pooya, Safoora Yousefi, Mohamed Amgad, David A. Gutman, Jill S. Barnholtz-Sloan, José E. Velázquez Vega, Daniel J. Brat, and Lee A. D. Cooper. 2018. "Predicting Cancer Outcomes from Histology and Genomics Using Convolutional Networks." *Proceedings of the National Academy of Sciences* 115(13): E2970–E2979. doi:10.1073/pnas.1717139115

Paparrizos, John, Ryen W. White, and Eric Horvitz. 2016. "Screening for Pancreatic Adenocarcinoma Using Signals from Web Search Logs: Feasibility Study and Results." *Journal of Oncology Practice* 12 (8): 737–744. doi:10.1200/JOP.2015.010504

Rajkomar, Alvin, Jeffrey Dean, and Isaac Kohane. 2019. "Machine Learning in Medicine." *New England Journal of Medicine*, April. Massachusetts Medical Society. www.nejm.org/doi/10.1056/NEJMra1814259.

Sahoo, Prasan Kumar, Hiren Kumar Thakkar, and Ming-Yih Lee. 2017. "A Cardiac Early Warning System with Multi Channel SCG and ECG Monitoring for Mobile Health." *Sensors* (Basel, Switzerland) 17(4). doi:10.3390/s17040711

Shickel, Benjamin, Patrick Tighe, Azra Bihorac, and Parisa Rashidi. 2018. "Deep EHR: A Survey of Recent Advances in Deep Learning Techniques for Electronic Health Record (EHR) Analysis." *IEEE Journal of Biomedical and Health Informatics* 22(5): 1589–1604. doi:10.1109/JBHI.2017.2767063

Shilo, Smadar, Hagai Rossman, and Eran Segal. 2020. "Axes of a Revolution: Challenges and Promises of Big Data in Healthcare." *Nature Medicine* 26 (1): 29–38. doi:10.1038/s41591-019-0727-5

Silver, David, Thomas Hubert, Julian Schrittwieser, Ioannis Antonoglou, Matthew Lai, Arthur Guez, Marc Lanctot, et al. 2018. "A General Reinforcement Learning Algorithm That Masters Chess, Shogi, and Go through Self-Play." *Science* 362 (6419): 1140–1144. doi:10.1126/science.aar6404

Stokes, Jonathan M., Kevin Yang, Kyle Swanson, Wengong Jin, Andres Cubillos-Ruiz, Nina M. Donghia, Craig R. MacNair, et al. 2020. "A Deep Learning Approach to Antibiotic Discovery." *Cell* 180(4): 688–702.e13. doi:10.1016/j.cell.2020.01.021

Tsherniak, Aviad, Francisca Vazquez, Phil G. Montgomery, Barbara A. Weir, Gregory Kryukov, Glenn S. Cowley, Stanley Gill, et al. 2017. "Defining a Cancer Dependency Map." *Cell* 170 (3): 564–576.e16. doi:10.1016/j.cell.2017.06.010

Venter, J. C., M. D. Adams, E. W. Myers, P. W. Li, R. J. Mural, G. G. Sutton, H. O. Smith, et al. 2001. "The Sequence of the Human Genome." *Science (New York, N.Y.)* 291(5507): 1304–1351. doi:10.1126/science.1058040

Vermeulen, Roel, Emma L. Schymanski, Albert-László Barabási, and Gary W. Miller. 2020. "The Exposome and Health: Where Chemistry Meets Biology." *Science* 367 (6476): 392–396. doi:10.1126/science.aay3164

Way, Gregory P., and Casey S. Greene. 2018. "Discovering Pathway and Cell-Type Signatures in Transcriptomic Compendia with Machine Learning." e27229v1. PeerJ Inc. doi:10.7287/peerj.preprints.27229v1

White, Ryen W., and Eric Horvitz. 2017. "Evaluation of the Feasibility of Screening Patients for Early Signs of Lung Carcinoma in Web Search Logs." *JAMA Oncology* 3(3): 398–401. doi:10.1001/jamaoncol.2016.4911

Xiong, Hui Y., Babak Alipanahi, Leo J. Lee, Hannes Bretschneider, Daniele Merico, Ryan K. C. Yuen, Yimin Hua, et al. 2015. "RNA Splicing. The Human Splicing Code Reveals New Insights into the Genetic Determinants of Disease." *Science (New York, N.Y.)* 347 (6218): 1254806. doi:10.1126/science.1254806

Yang, Qiang, Yang Liu, Tianjian Chen, and Yongxin Tong. 2019. "Federated Machine Learning: Concept and Applications." *ACM Transactions on Intelligent Systems and Technology* 10(2): 12:1–12:19. doi:10.1145/3298981

Zhang, Zhiqiang, Yi Zhao, Xiangke Liao, Wenqiang Shi, Kenli Li, Quan Zou, and Shaoliang Peng. 2019. "Deep Learning in Omics: A Survey and Guideline." *Briefings in Functional Genomics* 18 (1): 41–57. doi:10.1093/bfgp/ely030

Zitnik, Marinka, Francis Nguyen, Bo Wang, Jure Leskovec, Anna Goldenberg, and Michael M. Hoffman. 2019. "Machine Learning for Integrating Data in Biology and Medicine: Principles, Practice, and Opportunities." *Information Fusion* 50 (October): 71–91. doi:10.1016/j.inffus.2018.09.012

2

Predictive Analysis for Flood Risk Mapping Utilizing Machine Learning Approach

Aditya Singh,[1] Sunil Khatri,[2] Sandhya Save,[3] and Hemant Kasturiwale[4]

[1]*Department of Electronics Engineering, Thakur College of Engineering & Technology, Kandivali, India*

[2]*Assistant Professor, Department of Electronics Engineering, Thakur College of Engineering & Technology, Kandivali, India*

[3]*Professor, Department of Electronics Engineering, Thakur College of Engineering & Technology, Kandivali, India*

[4]*Associate Professor, Department of Electronics Engineering, Thakur College of Engineering & Technology, Kandivali, India*

CONTENTS

DOI: 10.1201/9781003104858-2

2.1   Introduction

Land cover refers to the surface cover of the ground, whether it is vegetation, water, or bare soil. In short, land cover indicates the physical type of the land (water, snow, grassland, soil). Land use refers to human activities that are directly related to the land (agricultural land, canal, built-up land, and other human-made characteristics). Together, they form a pattern called as land use and land cover (LULC) pattern. This pattern is an outcome of socioeconomic and natural factors and their utilization by humans in time and space [1]. Land use and land cover mapping is carried out to study the land utilization pattern and future planning and management of the land resource to either avoid economic loss due to natural factors or improve the ecological balance in the system. Also, it is used for planning and development of land parcel. The assessment of risks and the development of risk maps for future land use and infrastructure development is essential. Therefore, the change in land use land cover (LULC) pattern is detected. This pattern is affected by floods that occur in a particular region. Floods can cause devastation to human lives, property, and possessions as well as disruptions in communications. Flood tends to occur when rainfall is very high, absorption is very low, and when overflows are not controllable. Some other factors responsible for floods are unconditional rainfall, increase in the number of low-lying areas, rising sea level, poor sewage system, and low absorption capability of soil. Using predictive analysis and remote sensing methods, floods can be predicted in advance, which will not only help in detecting imminent patterns but will also help us in preparing beforehand to tackle such situations and come up with proper contingency plans. We plan to model and predict future impacts as well as the rate of occurrence of floods using image processing and predictive analysis.

Earlier, when no remote sensed data and computer assistance were available, land use/land cover changes were detected with the aid of tracing papers and topographical sheets. But this method was tedious, inefficient, and inaccurate. Studying large areas required considerable amount of effort and time. Conventional ground methods used for land use mapping are labor-intensive, time-consuming, and are done less frequently [2]. Thus, with the advent of satellite remote sensing techniques, which allow for easy collection of data over a variety of scales and resolutions, preparing accurate land use land cover maps was feasible. Also, monitoring changes at regular intervals of time became relatively simpler. In case of an inaccessible region, the only method of acquiring required data is by applying this technique. Today, remote sensing and geographical information system (GIS) technology has enabled ecologists and natural resources managers to acquire timely data and observe periodic changes. Predicting floods in any location has remained a longstanding challenge, which plays a major role in emergency management.

Remote sensing makes it easier to locate floods that have spread over a large region, thereby making it easier to plan rescue missions quickly. Also, remote sensing is a relatively cheap and efficient method in comparison to traditional methods. Predictive analytics offers a unique opportunity to identify certain trends and patterns that are used to identify future outcomes. Implementation of predictive mapping techniques became easier with the advent of predictive modeling techniques and their simple incorporation with other technologies. Thus, by combining both remote sensing and predictive analysis, a model can be created that will help in the detection of floods well in advance.

High-definition satellite images can be integrated with socioeconomic data in order to build environmental, economic, and social threat prediction models. Using these models, one can build demand-driven applications to help public and private organizations understand, prepare, and respond to economic, social, and humanitarian losses in a timely manner. Such models can be used to make near-real-time remote sensing applications by adding additional layers of response tools, like alerts and advisories. Ideally, these technologies would be able to be scaled up from local to global for emergency management and risk management. Some of the benefits it can offer are disaster risk reduction, event prediction for timely response and recovery, and allowing stakeholders to better target investments and protect assets. With the help of this study, governments can formulate suitable policies that are in the interest of the people and which will also help us in attaining ecological stability.

Among the cataclysmic events, floods are the most dangerous in causing severe harm to human life, foundation, farming, and the financial frameworks [1]. Flood risk analysis is undertaken to research land use and is also used in the planning and management of future land resources, which is done either to prevent economic loss due to natural causes or to enhance the system's ecological balance. Using predictive analysis and remote sensing methods, floods can be predicted in advance, which will not only help in detecting imminent patterns but will also help us in preparing beforehand to tackle such situations and come up with proper contingency plans. Implementation of predictive mapping techniques became simpler with the advent of predictive modeling techniques and its simple incorporation with other technologies [1]. Robust and precise prediction significantly leads to water resource management strategies, policy recommendations and research, as well as further forecasting and evacuation. Governments are actually under pressure to set up dependable and precise maps of flood risk regions and further prepare for effective management of floods [1]. Governments will be able to devise suitable policies with the help of this study that are in people's interest and which will eventually help to achieve ecological stability.

2.1.1   Machine Learning (ML)

A data-driven model is a type of model that uses ML techniques to build a relationship between input and output data, without having to worry too much about the underlying process. ML is a part of artificial intelligence (AI) process for inducing regularities and patterns, making implementation simpler, with low computation costs, quick training, validation, testing and evaluation, high performance compared to physical models, and comparatively less complex compared to physical models. One of the characteristics of ML algorithms that must be considered is that these models are a good as their training where the system, based on past data, learns the target tasks [1].

If the data is sparse or does not cover a variety of tasks, their learning will be limited, and therefore they will be unable to perform well when put into work.

2.1.2   ML Algorithms Used for Flood Risk Mapping

2.1.2.1   Artificial Neural Networks

Artificial neural networks is a model of computation based on biological neural network structure and functions. They are considered as a nonlinear modeling tool for statistical data where the complex associations between inputs and outputs are designed or patterns are associated [1]. As one of the most popular learning algorithms, ANNs are known to be flexible and efficient in modeling complex flood processes with large fault tolerance and accurate approximations, among the ML methods. ANN utilizes historical data and is more reliably used for flood prediction. ANN offers the most appropriate modeling technique and acceptable generalization capacity and speed compared to other models. Every ANN layer essentially carries out nonlinear input transformations from one vector space to another. The relatively subpar accuracy, the urge for iterative parameter adjustment, and the sluggish response to gradient-based learning process are major drawbacks to using ANN.

2.1.2.2   Multilayer Perception

For a more sophisticated modeling approach, the knack percolation learning algorithm calculates the propagation error in hidden network nodes individually. Nonlinear activation functions resolve the major drawbacks of linear activation function by allowing backpropagation because they have an input-related derivative function [1]. And they allow various layers of neurons to stack to create a deep neural network. Because of many variables, MLP is considered to be tougher to optimize.

2.1.2.3 Adaptive Neuro Fuzzy Inference System (ANFIS)

Fuzzy logic as a simplified mathematical model works to incorporate the knowledge of the expert into a fluid inference system. Fuzzy interference system (FIS) further imitates human learning through a less complex approximation feature, which offers great potential for nonlinear designing of extreme hydrological occurrences such as floods. Neurofuzzy has a back-propagation neural network (BPNN) variant and the least square test of error, quick and easy to implement, effective learning, and good generalization skills [1].

2.1.2.4 Wavelet Neuro Networks (WNN)

Wavelets transform (WT) can be used to analyze local time-series variation to extract information from different data sources. To boost data quality, it supports accurate decomposition of an original time sequence. This incorporates wavelet theory and the neural network (NN) in one theory. WNN, which incorporates WT and FFNN, and the wavelet-based regression model, which integrates WT and multiple linear regression (LR), were used to forecast floods in time series [1].

2.1.2.5 Support Vector Machine (SVM)

SVM is a supervised learning machine that functions on the theory of statistical learning and the rule of structural risk minimization. With a huge popularity among hydrologists for flood prediction, SVM and SVR were used as the alternative ML methods to ANNs [1]. It is ideal for linear and nonlinear classification as well as efficient input mapping into feature spaces. However, a high cost of computation and disappointing performance make it a lesser preferred option.

Table 2.1 presents a survey of the literature on work done by earlier researchers on similar tracks.

2.2 Methodology

2.2.1 Study Area

Mumbai is the capital of the state of Maharashtra, one of states with the highest population. It is located on a peninsula on the western coast of India sharing borders with Arabian Sea to the south and west, Mira Bhayander and Thane to the north and Navi Mumbai to the east [2]. The study area, that is, Mumbai suburban district, has an area of 446 sq. km. and a population

TABLE 2.1

Literature Review of Work Done by Various Authors over Similar Tracks

Presented at	Year	Name	Author	Inference
Journal of Environmental Protection and Ecology	2018	Remote sensing and geographic information system in the management of agricultural risks related to climate change	M. T. Esetlili, Y. Kurucu, G. Cicek, O. Demirtas	Prediction of probable reduction in the yields that might occur through monitoring and assessment.
International Journal of Remote Sensing	2009	Analysis of urban growth pattern using remote sensing and GIS	B. Bhatta	Study of urban growth pattern
University of Delhi, New Delhi	2018	Linear regression analysis and study	Khushbhu Kumari, Suniti Yadav	The concept of correlation and linear regression was understood and the relationship between them was obtained.
G. Rudolph et al.	2008	Prediction using linear regression and Markov chains	AnabelaSimoes, Ernesto Costa	Understanding of Markov chains and dynamic environment prediction
Landscape and Urban Planning 55	2001	Predicting land-cover and land-use change in the urban fringe	Erna Lo´peza, Gerardo Boccoa, Manuel Mendozaa, Emilio Duhaub	Markov chain implementation for prediction of change in land features
Computer Networks 33	2000	Link prediction and path analysis using Markov chains	Ramesh R. Sarukkai	Understanding fundamentals of Markov chain algorithms
The R Journal	2017	RQGIS: Integrating R with QGIS for statistical geo-computing	JannesMuenchow, PatrikSchratz, Alexander Brenning	Understanding the R package RQGIS
Journal of Geomatics	2019	Spatial pattern of urban growth using remote sensing and landscape metrics	Akintunde John Akinrinola	Seeks to critically understand the pattern of urban growth
Sustainability	2017	Land use/ land cover change modeling and the prediction of subsequent changes in ecosystem service values in a coastal area of China	Wei Wu, Eshetu Yirsaw, Xiaoping Shi, Habtamu Temesgen, Belew Bekele	Explores a future LULC simulation using a CA-Markov model

TABLE 2.1 (Continued)
Literature Review of Work Done by Various Authors over Similar Tracks

Presented at	Year	Name	Author	Inference
National Institute of Technology, Rourkela	2011	Land use and land cover change detection at Sukinda Valley using remote sensing and GIS	Biswajit Majumdar	Mapping of land use and land cover pattern
Journal of Applied Science Environment Management	2018	Land use/land cover change analysis using Markov-based model for Eleyele Reservoir	Bello Ho	Markov based LULC changes with respect to water bodies
International Multidisciplinary Scientific Geo Conference	2018	Accuracy analysis of the inland waters detection	Marina Gudelj, Mateo Gasparovic, MladenZrinjski	Workflow for this project
Landscape and Environment 10	2016	Specific features of NDVI, NDWI and MNDWI as Reflected in LC Categories	Szilard Szabo, Zoltan Gasci, BoglarkaBalazs	NDWI, NDVi, MNDWI coverages for water bodies LULC study

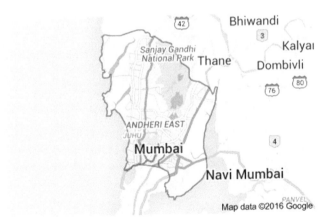

FIGURE 2.1
Geographical Region of Mumbai Suburban Region.

of 9,356,962 according to the 2011 census [3]. Thus, the density of Mumbai suburban district is 20,980 people per square kilometer. With elevations ranging from 10 meters to 15 meters, the city generally lies just above the sea level with an average elevation of 14 meters [4, 5]. This makes the city highly prone to flooding and waterlogging. Figure 2.1 shows a geographical map of Mumbai.

2.2.2 Remote Sensing

Earlier, when no remote sensed data was available, changes in LULC patterns were detected using tracing papers and topographic sheets. But that approach has been slow, unsuccessful, and unreliable. Wide areas of study involved a considerable amount of time, effort, and cost [5, 6]. Conventional solutions are labor-intensive, time-consuming, and less often performed. Therefore, the preparation of detailed land use and land cover maps and tracking adjustments at regular intervals of time is comparatively simpler with the advent of satellite remote sensing techniques [7]. In the event of inaccessible areas, the only method of data acquisition is by applying this technique. Quantum geographical information system (QGIS) is the software tool used for remote sensing purposes.

2.2.3 LULC Map Creation

LISS—III images were used in the study marked in the years 2011 to 2015. The processing of the image and classification was carried out using QGIS version 3.3 Zanzibar [8, 9]. Normalized difference water index (NDWI) is used to monitor changes related to water content in water bodies, using green and NIR wavelength.

$$NDWI = (NIR - SWIR)/(NIR + SWIR)$$

The Bands of the LISS-III Satellite and their Wavelengths are shown in Table 2.2.

2.2.4 State Selection

Since the Markov model is a stochastic state-based transition model, selection of states is an important step. The states need to be clearly distinguishable from the available LULC maps as that will provide greater accuracy [10].

TABLE 2.2

Bands of LISS-III Satellite and Their Respective Wavelengths

Band	Wavelength	Revisit time
BAND 2(VIS)	0.52 to 0.59	24
BAND 3(VIS)	0.62 to 0.68	24
BAND 4(NIR)	0.77 to 0.86	24
BAND 5(SWIR)	1.55 to 1.75	24

2.2.5 Markov Model

We have seen that whenever a succession of chance analyses frames a free preliminaries process, the potential results for each examination are equivalent and happen with a similar likelihood [11, 12]. Further, information on the results of the past investigations doesn't impact our expectations for the results of the following test. The circulation for the results of a solitary trial is adequate to develop a tree and a tree measure for a grouping of n trials, and we can respond to any likelihood question about these trials by utilizing this tree measure. Current likelihood hypothesis reads chance procedures for which the information of past results impacts expectations for future analyses. At a fundamental level, whenever we watch a grouping of chance analysis, the entirety of the past results could impact our forecasts for the following test [6, 12].

Markov chain calculates how much land is estimated to change from the latest date to the expected date [7]. In this method, the transfer probabilities are the output, which is a matrix that records the likelihood that each land cover class will move to each other class[9, 12]. Through the Markov chain simulation, the analysis of two different dates of the LULC images creates the transition matrices, a transition area matrix, and a set of conditional probability images. The transition probability matrix for the period of 2008–2015 was determined to predict the 2019 LULC map and thus any future given date for prediction [13]. The probability of one pixel moving to another LULC or staying in the original LULC can be calculated by producing a transition matrix of probabilities. The transition matrix is generally important for predicting a future classification map [14]. If the hypothesis tested has no significant differences in the observed LULC and the predicted LULC, the model is considered successful for future predictions [15, 16].

2.2.6 Transition Matrix Calculations

Figure 2.2 shows State Markov Model. LISS—III images are used for remote sensing analysis. This part includes preprocessing of images followed by

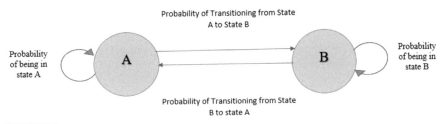

FIGURE 2.2
State Markov Model.

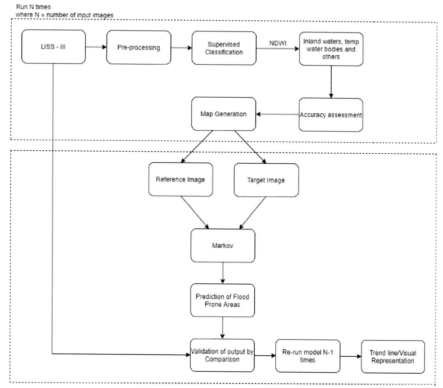

FIGURE 2.3
Block Diagram of the proposed model.

supervised classification. NDWI is calculated in this block. After accuracy assessment, the corresponding maps are generated [17]. The Markov model depends and makes transitions between a fixed number of states. The states are classified as transient and absorbing states [18]. The absorbing states mark an end to the process and thus need to be selected accordingly. The initial state transition probabilities are determined based on the basis of the initial land cover of a specific state [19]. The transition matrix can be determined by a change in the first iteration. This determines the number of states or in real world, the time required by the process to end up in an absorbing state [20]. This will obviously exist for all states since the given problem is an absorbing state problem, but we will estimate the transition to the absorption state in a finite amount of time. Multiple iterations will be carried out and finally we will be able to see the trends or pattern this model will follow [21, 22]. Figure 2.3 shows Block Diagram of the proposed model.

FIGURE 2.4
NDWI Map of the year 2011.

2.3   Results

Using QGIS software as a tool for remote sensing, the study area was analyzed for a period of five years starting from 2011 to 2015. During this analysis, NDWI values were calculated for each year and their corresponding maps were generated. Figure 2.4 shows NDWI Map of the year 2011. The NDWI maps and rainfall data for year 2011–2015 are shown in Figure 2.5. In order to have a clear picture of the water bodies present in the study area, Band 5 of LISS—III satellite is used for short-wave infrared sensing. These NDWI values, along with rainfall data in the above-mentioned years, are taken as parameters or variables for carrying out predictive analysis using the Markov chain model. Figure 2.6 shows NDWI Map of the year 2015 whereas Figure 2.7 shows Rainfall Data of the year 2015.

The Markov model, a stochastic state-based transition model, is used with two different states. The states need to be clearly distinguishable from the available LULC maps as it will provide greater accuracy. In this case, two states are decided. The first state is the state of being in flood (A) and the next

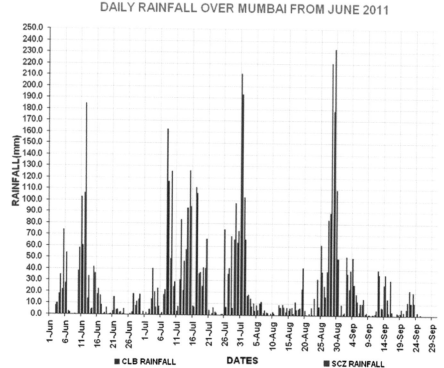

FIGURE 2.5
Rainfall Data of the year 2011.

state is the state of not being in the flood (B). So, basically there are two states, viz. A and B. The transition probabilities will be calculated and the transition matrix will be formed depending upon these states. The transition matrix calculated for years 2011–2015 is shown in Figure 2.8.

Figure 2.9 shows the implementation of two-state Markov model using the MATLAB simulation tool. The time steps for simulation are 5, and the transition probabilities of each element are depicted in this simulation.

2.4   Conclusion

The geographic analysis of study area, viz. coverage area, population, and height from sea, is performed to make a prediction about the possibility of flooding and waterlogging. Superior quality satellite images are integrated

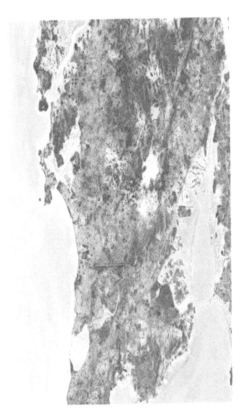

FIGURE 2.6
NDWI Map of the year 2015.

with socioeconomic data in order to build environmental, economic, and social threat prediction models. The processing of the image and classification is carried out using QGIS version 3.3 Zanzibar. The processing of images is done on basis of calculation of normalized difference water NDWI values, which are related to water content in water bodies, using green and NIR wavelengths. Markov chains provide a stochastic model of dispersion that applies to singular particles, which are used for stochastic dissemination we've examined. This stochastic dissemination likewise gives a valuable model of the spread of data all through the picture. This Markov chain-based model gives an accurate prediction of flood conditions.

Such models can be used to make near-real-time remote sensing applications by adding additional layers of response tools, like alerts and advisories, for disaster and risk management. More training data can be used to improve the accuracy of this designed model.

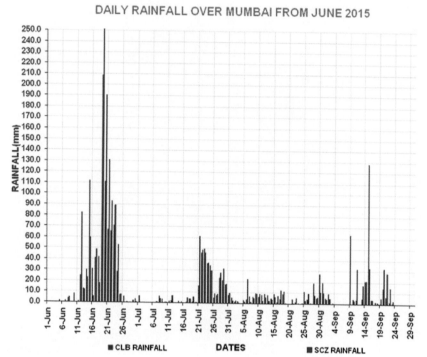

FIGURE 2.7
Rainfall Data of the year 2015.

2011:

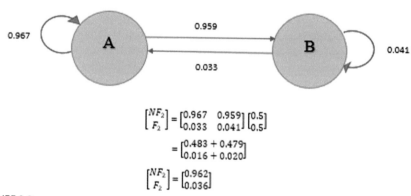

$$\begin{bmatrix} NF_2 \\ F_2 \end{bmatrix} = \begin{bmatrix} 0.967 & 0.959 \\ 0.033 & 0.041 \end{bmatrix} \begin{bmatrix} 0.5 \\ 0.5 \end{bmatrix}$$

$$= \begin{bmatrix} 0.483 + 0.479 \\ 0.016 + 0.020 \end{bmatrix}$$

$$\begin{bmatrix} NF_2 \\ F_2 \end{bmatrix} = \begin{bmatrix} 0.962 \\ 0.036 \end{bmatrix}$$

FIGURE 2.8a
Two State Markov Models for the year 2011–2015.

2015:

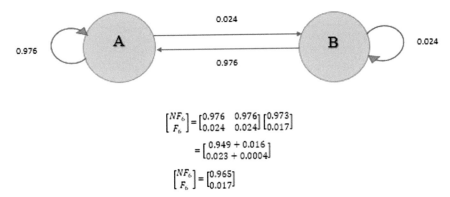

$$\begin{bmatrix} NF_6 \\ F_6 \end{bmatrix} = \begin{bmatrix} 0.976 & 0.976 \\ 0.024 & 0.024 \end{bmatrix} \begin{bmatrix} 0.973 \\ 0.017 \end{bmatrix}$$

$$= \begin{bmatrix} 0.949 + 0.016 \\ 0.023 + 0.0004 \end{bmatrix}$$

$$\begin{bmatrix} NF_6 \\ F_6 \end{bmatrix} = \begin{bmatrix} 0.965 \\ 0.017 \end{bmatrix}$$

FIGURE 2.8b
Two State Markov Models for the year 2011–2015.

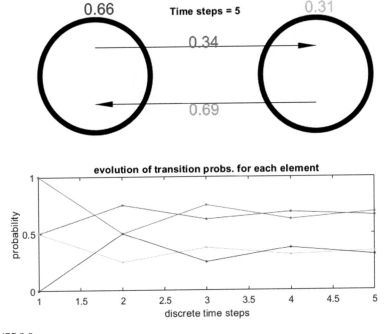

FIGURE 2.9
State Markov Model Implementation on MATLAB.

References

[1] Mosavi, A.; Ozturk, P.; Chau, K.-W. Flood Prediction Using Machine Learning Models: Literature Review. Water 2018, 10, 1536.

[2] www.census2011.co.in/census/district/357-mumbai-city.html

[3] http://pibmumbai.gov.in/English/PDF/E2013_PR798.PDF

[4] Biswajit Majumdar, "Land use and land cover change detection at Sukinda Valley using remote sensing and GIS", National Institute of Technology Rourkela, 2011.

[5] Rahel Hamad, Heiko Balzter, Kamal Kolo, "Predicting land use/land cover changes using a CA-Markov model under two different scenarios", Sustainability, 2018.

[6] Jannes Muenchow, Patrik Schratz, Alexander Brenning, "RQGIS: Integrating R with QGIS for statistical computing," R Journal, 2017.

[7] S Junaida, Sulaiman and Siti Hajar, Wahab (2018) Heavy Rainfall Forecasting Model Using Artificial Neural Network for Flood Prone Area. In: IT Convergence and Security 2017. Lecture Notes in Electrical Engineering, 449 . Springer, Singapore, pp. 68–76.

[8] S. Khatri and H. Kasturiwale, "Quality assessment of Median filtering techniques for impulse noise removal from digital images," 2016 3rd International Conference on Advanced Computing and Communication Systems (ICACCS), 2016, pp. 1–4, doi: 10.1109/ICACCS.2016.7586331.

[9] J. Abbot,; J. Marohasy, "Input selection and optimization for monthly rainfall forecasting in Queensland, Australia, using artificial neural networks. Atmospheric Research, pp. 166–178, 2014.

[10] A. Jain,, S. Prasad Indurthy, Closure to "comparative analysis of event-based rainfall-runoff modeling techniques—deterministic, statistical, and artificial neural networks".. Journal of Hydrologic Engineering, vol. 9, pp. 551–553, 2004.

[11] R.C. Deo, M. Sahin, Application of the artificial neural network model for prediction of monthly standardized precipitation and evapotranspiration index using hydrometeorological parameters and climate indices in eastern Australia. Atmospheric Research, pp. 65–81, 2015.

[12] M.K. Tiwari, C. Chatterjee, Development of an accurate and reliable hourly flood forecasting model using wavelet–bootstrap– ANN (WBANNss) hybrid approach. Journal of Hydrology, pp. 458–470, 2010.

[13] C.A. Guimarães Santos, G.B.L.d. Silva. Daily streamflow forecasting using a wavelet transform and artificial neural network hybrid models. Hydrological Sciences Journal, pp. 312–324, 2014.

[14] S. Supratid, T. Aribarg, S. Supharatid. An integration of stationary wavelet transform and nonlinear autoregressive neural network with exogenous input for baseline and future forecasting of reservoir inflow. Water Resources Management, vol. 31, pp. 4023–4043, 2017.

[15] E. Dubossarsky, J.H. Friedman, J.T. Ormerod,. M.P. Wand. Wavelet-based gradient boosting. Statistics and Computing, vol. 26, pp. 93–105, 2016.

[16] K. Kasiviswanathan, J. He,; K. Sudheer, J.-H. Tay. Potential application of wavelet neural network ensemble to forecast streamflow for flood management. Journal of Hydrology, pp. 161–173, 2016.

[17] L.A. Zadeh. Soft computing and fuzzy logic. In *Fuzzy Sets, Fuzzy Logic, and Fuzzy Systems: Selected Papers by Lotfi a Zadeh*, World Scientific: pp. 796–804, 1996.

[18] P.S. Yu, T.C. Yang, S.Y. Chen, C.M. Kuo, H.W. Tseng.Comparison of random forests and support vector machine for real-time radar-derived rainfall forecasting. Journal of Hydrology, vol. 552, pp. 92–104, 2017.

[19] S. Li, K. Ma, Z. Jin, Y. Zhu. In A new flood forecasting model based on svm and boosting learning algorithms, Evolutionary Computation (CEC), IEEE Conference, pp. 1343–1348, 2016.

[20] E. Danso-Amoako, M. Scholz, N. Kalimeris, Q. Yang, J. Shao. Predicting dam failure risk for sustainable flood retention basins: A generic case study for the wider greater Manchester area. Computers, Environment and Urban Systemsvol Vol. 36, pp. 423–433, 2012.

[21] Y.B. Dibike, S. Velickov, D. Solomatine, M.B. Abbott. Model induction with support vector machines: Introduction and applications. Journal of Computing in Civil Engineering, vol.15, pp. 208–216, 2001.

[22] Rahel Hamad, Heiko Balzter, Kamal Kolo, Predicting land use/land cover changes using a CA-Markov model under two different scenarios, 2018.

3

Machine Learning for Risk Analysis

Parita Jain,[1,*] Puneet Kumar Aggarwal,[2] Kshirja Makar,[3] Riya Garg,[3]
Jaya Mehta,[3] and Poorvi Chaudhary[3]

[1]KIET Group of Institutions, UP, India

[2]ABES Engineering College, UP, India

[3]HMRITM, Delhi, India

[*]paritajain23@gmail.com

CONTENTS

3.1 Introduction

Machine learning (ML) is the new development in technology. It has the proficiency of replacing emphatic programming of the devices. It is generally hinged on the conception that a substantial amount of data is provided and on the basis of that data and some algorithms, the machine is trained

DOI: 10.1201/9781003104858-3

and different machine modules are fabricated. The decisions made by the machines on the basis of data provided are immensely efficient and accurate. With increased use of technology, ingenious crimes and risks are also escalating. In the modern world, the most important constituent to take off is risk management and its intensifying production. This chapter provides all the information regarding how machine learning has been applied in risk assessment (Apostolakis, 2004; Aven, 2012).

Large-scale organizations, companies, and institutions are prone to risks like frauds; consequently, they are sticking to various machine learning techniques that can prevent or abate these frauds or risks. The chapter is an amalgamation of various strategies for changing the perspective of risk assessment. On the basis of risk assessment with machine learning, various case studies covering different aspects are presented in the chapter. These case studies effectively portray the seriousness of risk assessment. Case studies elucidate constructing techniques to diminish risks and act as a reference for future. Several applications are also incorporated in this chapter, which lead us to the significance of risk assessment in several industries and organizations for protection from a large amount of frauds and deceptions altogether with the integration of machine learning in this peculiar field (Chen, 2008; Cheng, 2016).

The essence of this chapter is related to the arguments that provide a solution for risks found in the literature followed by a conclusion, which covers machine learning in the field of risk management as a whole.

3.1.1 Machine Learning

Machine learning can be described as a subcategory of Artificial Intelligence (AI), which has its main focus on examining and recognizing patterns and arrangements in data to facilitate features such as training, thinking, decision making, learning, and researching without interference. Machine learning allows the user to pack an enormous sum of data with a computer algorithm and allows the computer to examine and analyze the data to make recommendations based on the input received. If some features require redesigning, they are classified and improved for a better design for the future (Creedy, 2011; Comfort, 2019).

The main aim of the technology is to produce an easy-to-use mechanism, which works on making decisions by the computational algorithm. Variables, algorithms, and innovations are accountable for making decisions. Awareness toward the solution is needed for the better learning of the working of the systems and thus helps in understanding the path to reach the result.

In the initial stages of implementation of the algorithm, input or data is fed to the machine provided the result is already known for that set of feed data. The changes are then made to produce the result (Diekmann, 1992; Durga,

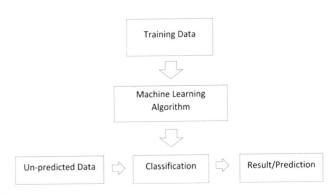

FIGURE 3.1
Machine learning algorithm process.

2009; Goodfellow, 2016). The efficiency of the result depends on the amount and the quantity of the data that is fed to the machine as shown in Figure 3.1.

3.1.2 Risk Analysis

Risk is referred to as the occurrence of undesired events while running a project, which creates a negative impact on the achievement of the goals of that project. Risk is the possibility of an unwanted or harmful event. Risk analysis is done by adopting different methods in order to check the probability of the happening of the risk and it is performed in order to remove the chances or minimize the probability of the occurrence of the harmful events. Risk analysis is desirable as the occurrence of risk is directly proportional to the amount of losses faced in a project. Every technology or project has an equal chance of being successful as well as becoming a failure and hence the analysis is done to reduce this chance, which is called the risk of failure (Kaplan 1981; Hastie, 2009; Hauge, 2015; Haugen, 2015; Khakzad, 2015).

Risk analysis includes identifying the type of hazard that can be associated with the project, which may be a chemical hazard, mechanical hazard, or even technical hazard. It is a process that is analytical in nature and the aim is to know all the desired information related to undesirable events. It involves an analysis of the hazards that have occurred in the past and also which have the probability of occurring in future (Khakzad, 2013a; 2013b). It not only analyzes the hazards but also their consequences on the system so that appropriate measures can be taken. The main objective is to increase the chances of success and at the same time minimize the cost or investment on the project. As important risk analysis is, it is even more difficult to be performed. The traditional method of using employees requires a lot of time to complete the process (King, 2001; Kongsvik, 2015; Landucci, 2016a, 2016b). But now with the growth of industries, the amount of data generated is very high, operations performed on a project are more than before, but the time to complete that project is lower. So, in order to fulfill the demand versus time ratio, industries use technical and more reliable methods such as ML. There are different methods in which risk analysis can be done. These methods are as follows:

(i) *Qualitative methods:* It is a method generally used in the process of decision making or before decision making. It involves various predictions based on previous experiences; different judgments are passed by different members of the project. This is not a very accurate method and is generally done at the initial stage of the project (Jain, 2018; Jain, 2019). The major aim is to improve the quality of the project based on past experiences. This method is suggested when there is no time constraint and the risk level is quite low. This method does not make use of any algorithms or numerical data. It is performed through brainstorming, by questionnaire, or interviews.

(ii) *Quantitative methods:* This method is way more numerical and reliable than the qualitative methods. In this method, different numerical values can be assigned to different risks identified based on the probability of their occurrences and the algorithms are performed to calculate the level of the risk of the project (Aggarwal 2018; Aggarwal 2019; Jain 2020). In this method, we do not assume what is going to happen. It involves analyzing the probability of the number of occurrences, analysis of the consequences as well as the after-effects of the solution. This method is a little expensive as it requires machines and computer systems, but it is very accurate in results.

(iii) *Semiquantitative methods:* These methods are neither fully qualitative nor completely numerical or algorithm based. This is a mid-way method adapted according to the needs of the project and clients. The method used depends on the level of risk of the project. In this method, various projects are classified as low-, middle-, or high-risk level projects. This method is used in small-scale projects or where the client is not sure about the requirements and hence is not able to predict the exact level of the associated risk.

The major components of risk analysis include hazard identification, risk assessment, risk management, and risk communication (Nivolianitou, 2004; Nobre, 2009; Musgrave, 2013; Lasi, 2014). The step-by-step process is shown in Figure 3.2.

(i) *Hazard identification:* The primary step to proceed toward risk analysis is the identification of the problems, risks, or hazards that can be faced in the future. The number of hazards determines the level of risk for the project. More the number of hazards higher is the level of risk. Once the

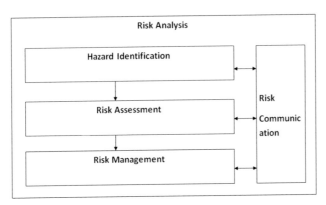

FIGURE 3.2
Components of risk analysis.

identification is done, this stage is followed by risk perception, which means acceptance of the risk identified. How the risk is perceived varies from person to person and also upon the characteristics of the risk. If the client wants the project to be successful and earn a profit, then he/she is bound to accept the risk and also take appropriate measures in order to minimize it. In order to cure the problem first, we need to know the problem; similarly, in order to remove the risks first, the hazard needs to be identified as well as accepted and not ignored.

(ii) *Risk assessment:* When the level of risk increases, it becomes necessary to assess these risks in order to find solutions at the right time. This is done to ensure the safety of the project as well as the workers and users (Paltrinieri, 2015).

(iii) *Risk management:* This refers to managing the risks that are being assessed and analyzed. Different strategies are used to manage the risks. These strategies aim to remove the negative effects of the risk and increase productivity (Pasman, 2014).

(iv) *Risk communication:* This refers to making this process interactive between the clients, developers, and users by exchanging and transferring the information related to risk with each other. This stage involves the active participation of all those who are associated with a project directly or indirectly, that is, from the lower to higher level in order to make the process efficient and fulfill the demands of all stages. It is important because it makes the decision-making process easier as the decision involving the lowest risk probability is chosen. It is done by discussing and analyzing different hazards and their impact. It reduces the gap between the expectations and results provided. Also, it gives voice to the users to address their disappointments, which are also kind of a risk (Aven 2014; Bucelli, 2017).

3.2   Risk Assessment and Risk Management

When the level of risk increases, it becomes necessary to assess these risks in order to find the solutions at the right time. This is done to ensure the safety of the project as well as the workers and users. For example, mechanical project risks can be dangerous for workers, which are technology-related project risks that can prove to be dangerous for users' data. To assess the hazards, one must figure out the origin, consequences, and the results. To identify the origin, all the sources must be assessed to understand where things could go wrong. Then, it has to be figured out how much are these risks exposed to the outside environment. This stage is known as risk estimation in which individual systems are assessed to know the contributions of each source and functionality toward the risk (Oien, 2011; Noh, 2014; Nyvlt, 2015).

Understanding the individual systems helps in forming connections and links of the sources with the risk being produced during development. If the risk is assessed during the development process, it can be easily cured and removed (Nyvlt, 2012). It also helps to predict which features can produce failures in future and they can be continuously monitored and managed. Hence the qualitative analysis of risk assessment includes a detailed study of logical flow of data and the system functionalities.

Generally, the risks that are assessed are represented in the form of diagrams, flow charts, and decision trees for better understanding. For risk assessment, some factors such as the equipment used, the software developed, the functionalities produced, and target audience must be kept in mind. This step is very important in order to build the gap between the risk that is being predicted and that is being perceived (Paltrinieri, 2017). This is because the risk predicted is generally statistical in nature and is based on the theory of probability and number of assumptions. So, in order to bridge this gap, a quantitative and numerical approach must be adapted rather than estimating the factors on the basis of their frequency. This stage of analyzing the prediction with the result is called risk evaluation and is a part of risk assessment. Different tests are performed and various test drives are run to assess and predict the risk so that the final product is safe and successful. Also, various data sets are created in order to keep track of failures produced in the past and to ensure that same hazards are not produced in future.

3.2.1   Risk Management

The next step is to manage the risks that are being assessed and analyzed. Different strategies are used to manage the risks. These strategies aim to remove negative effects of the risk and increase productivity. Some people tend to manage the risk by avoiding it, but it is not the correct way to deal with the hazards. Because if they are not dealt with at the earlier stage, they will grow into something serious. So, the correct way of managing risks is first of all accepting them (Paltrinieri, 2019). Risk management is done on the basis of priority, which means the risk with highest probability of occurrence or the risk that can cause maximum amount of damage needs to be dealt with first and the risks that are expected to cause lower loss are prioritized in the descending order. Hence, a classification of the risks needs to be done on the basis of loss and probability.

Generally, the amount of loss faced is given more priority than the frequency of its occurrence. It is the most important step because if the risks are not managed properly, then there is no point in identifying and assessing them. The unmanaged risks reduce the rate of profit, durability, quality, and reliability of the project. It can also affect the brand value of a company and decrease the trust of users. The problem or limitation faced by this phase is the limited amount of technique and resources available to manage the hazards and the increase in cost due to risk management. The steps required in risk management are:

(i)   Identification of the risk and its resource and domain

(ii)   Considering the impact of this on the system

(iii) Probability of occurrence

(iv)   Impact on effective cost

(v)   Consequences on the project

(vi)   Classification on the basis of priority

(vii) Assessing the constraints that can be faced

(viii) Describing the needs of users and agenda for doing this activity

(ix)   Engaging in discussions and communication to decide the managing technique

(x)   Finally, developing an analytical approach in order to manage the risk

(xi)   Organizing the resources and cost required for the managing process

Thus, risk management is done on the basis of results obtained from risk assessment and then the appropriate techniques are chosen by managers.

3.3   Risks in the Business World

As we all know, any business, small or big, is prone to risk. Some of them even cause loss of profits to the extent of bankruptcy. The large firms have expanded their sections of risk management. The smaller firms or businesses are still not able to find systematic ways to handle the issue. Comprehensive business plans are required to have a successful business, but the plans need to be updated with time (Svozil, 1997). With changes in the market and its standards, the old methods of risk management and analysis require modification. The major risks that the firms face are the following:

(i) *Strategic risk:* As the name suggests, strategic risk is strategy-related, in which the planning is not appropriate for the successful development of the business. This is due to the technological changes in the society, new competitors with modern technologies entering the market, rapid changes in the customer demand section, and an increase in the cost of raw material. Studies show the example of firms that suffer strategic risk. Some trained their models to gain success, and some did not. An example of the same is Kodak, which had a decent position in the photography market. When engineers developed the digital camera, it was as a threat to its central marketing model, and Kodak failed to modify it. If Kodak had considered the strategic risk more thoroughly, it could have led to a better future for the company. Handling a strategic risk is not always destructive as there are examples of brands that were able to see it as a chance of change like Xerox. It became a success with a single and satisfactory product, that is, a photocopy machine. The company tried to add a laser printer to enhance the business model and was ready to face the statistical risk. The company saw the positive side of the change due to good planning.

(ii) *Compliance risk:* Any company always tries to increase their profits and business prospects, but they have to comply with law while doing the same. The laws constantly change with time and regulation is required. For example, a person makes cosmetic products in his factory and sells it across the country. The person is doing so well in the industry and thus wants to expand the business to another country. Now the business is under the compliance risk. The countries have some rules regarding the import of products and these rules and regulations may cost money and may not give a bigger margin in profits in comparison to the expected margins. The same condition may arise if the product line is changed.

(iii) *Operational risk:* This type of risk is due to the unexpected failure in the way the company operates. The risks include mechanical failure, technical failure, or failure due to people. If a person is to be given a salary of ₹10,000 and is handed a check for ₹1,000, then it is a failure of process and people. This may look like a small problem in the example but in the big industry, it can cause issues with bigger amounts. Other problems like natural disasters and power cuts can cause operational damage. From these examples, operational risk can be defined as any issue or risk that damages the function or operation of a business (Villa, 2016a, 2016b).

(iv) *Financial risk:* Most of the times, the risk that the firm faces is categorized as financial risk, that is, a risk having a financial issue leading to low revenue and decreased profits. Let us assume that a client has the credit extension of 30 days and is a large source of the company's revenue (Marchi, 1999). The company, in this case, is under financial risk as a great deal of trouble can be caused if the payment is not done on time. Similar is the case when a particular company is in great debt. International businesses undergo huge financial risks. We can go back to the example of cosmetics where the person wanted to sell the products in the international market and if the person tries to sell the goods in the United States, UK, France, and India, the company may have to bear the conversion charges of these different currencies. This type of example comes under financial risk.

(v) *Reputational risk:* One thing that is common to all the businesses is their reputation, be it a small size business or a multinational firm. If the reputation is positive, then the selling of commodity and recruitment of employees become easier. If the reputation is damaged, then there is an immediate loss in revenue and the employees might leave the company too. The advertising agencies may not show interest in the work related to the company and so on (Yang, 2015).

3.3.1 Models in the Field of Risk Management

Various models have been proposed for managing risks in different scenarios. Some of the models are explained as follows:

(i) *Interpretability:* Machine learning designs have a character of signifying results that is hard to understand or evaluate. To increase the interpretability, some models are used as follows:

    (a) *Linear monotonic models*: Linear coefficients act as a huge support for the exhibition of the output. Linear regression models come under this category.

    (b) *Nonlinear monotonic models*: Restraining data so that we can see either a falling or a rising global relationship where the variables are simplified for producing the result. Gradient boosting model is an example of the same.

    (c) *Nonlinear no monotonic model*: Methodologies such as local interpretable model-agnostic explanations or Shapley values help ensure local interpretability. The examples include models like unconstrained deep learning models.

(ii) *Bias:* This feature takes care of the trueness of the system. The validation takes care of the proper implementation of rules for fairness in the mechanism. Four approaches are generally used depending on the requirements:

    (a) *Demographic blindness*: Settlements are composed utilizing a restricted assortment of characteristics that are related to the preserved classes.

    (b) *Demographic parity*: For each protected class, results are proportionally the same.

(c) *Equal opportunity*: For each protected class, true positive standards are similar.

(d) *Equal odds*: For each protected class, true-positive and false-positive rates are equal.

(iii) *Feature engineering*: It is the method of managing data to create innovations that can be executed with the help of machine learning algorithms. It is the process of selecting significant characteristics from a fresh pool of data and converting them into forms that are fit for machine learning. Feature engineering model development in ML is complicated in comparison to the models used traditionally. The first reason for this is that machine learning models can combine a higher amount of information. The second is that there is a need for feature engineering that is required for disorganized data references at the pre-processing level before the training method can start.

(iv) *Hyper parameters:* The variables which define the system composition and decide the network training method are the hyper parameters. It is very important to understand the variables and determine the appropriate selection of hyper parameters. The approaches for the selection of hyper parameters include the latest practices used in the industry and expert judgment.

(v) *Production readiness:* Machine learning models, despite being algorithmic, require a lot of computation. This element is generally viewed in the process of model development. The validation is done previously to evaluate a variety of model risks connected with its usage and for machine learning, they expand their scope. There is a requirement of setting a limitation on the data flow through the model, keeping in mind the run time and the architecture of the model.

(vi) *Dynamic model calibration:* Sometimes, there is a dynamic change in the parameters in some types of models depicting the data patterns. The validators can easily decide which dynamic calibration is best suited for the firm. The factors that are evaluated include the development of a monitoring plan, ensuring proper control to reduce risk in accordance with the usage of the model.

3.4   Machine Learning Techniques for Risk Management and Assessment

Two types of ML techniques are used in risk analysis. In a supervised ML technique, the input is pre-decided under supervised conditions and generates an expected output (Wolpert, 2002; Jain 2020). In an unsupervised

ML technique, various kinds of inputs are given in undefined sets and an unexpected output is generated.

(i) *Supervised technique:* It is further divided into regression and classification. Regression technique is used to define the relationship between the variable assigned to the risks predicted during the assessment. For example, a regression equation for a credit-based assessment can be defined as the variable assigned to non-payment of the loan taken, which is further analyzed and sorted using a long range of variables that are independent and are assigned to functions such as employment condition of the person, past payment history, and property.

   This method is very useful as the independent variables facilitate in the management process and it is much better than the traditional method as it is self-explanatory and less theoretical and more practical. This method is suitable for big data sets as the variables that are not in use are deleted automatically. The other main technique is the classification technique. It is the most commonly used technique in which risks are prioritized and represented in the form of decision trees and data flow diagrams. This method is easy and more comprehensive due to the visual representation and hence requires less time in management.

   This model is understood by both technical and nontechnical members. The risks are branched into various categories according to the percentages, which indicate the probability of the occurrence of the various risks and hazards. This categorization and subdivision technique helps to determine the later consequences as well. This makes the sorting possible in less time as it contains only decision based yes/no type questions. The only limitation of this module is that its predictive power is low, which means it helps in analyzing but not in predicting when compared to the traditional methods.

(ii) *Unsupervised technique:* This method is further categorized into clustering. This is also based on subdivision, but it forms similar subgroups and does not predict any output. This is a kind of a grouping technique with no prediction function. The example of this technique is spam detection in emails. This means all the similar looking mails form a cluster that is if a mail is considered spam, then all the similar looking mails will be clustered as spam. This assignment of variables is lacking, which does not allow it to detect or predict anything. This just helps the analysts to figure out the patterns and similarities between the functions.

The ML techniques form various models that are quite complex in nature as it uses too many variables and other parameters in comparison of the output generated, which make it heavy as well as complex as shown in Figure 3.3. Due to the complexity in design, the interpretation time also increases, which

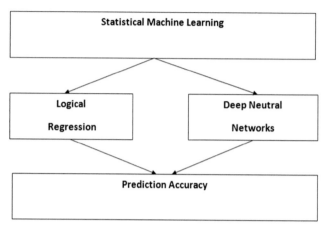

FIGURE 3.3
Machine learning assessment model techniques.

is undesirable. A good predictive model is generally hard to analyze. For efficient risk assessment and management, interpretation is as important as prediction. So, some methods can be applied on these models to overweight the observations in comparison to parameters or inputs so that more and more outputs can be generated. Such a method is called "boosting." Another way is running a model on different data sets many times so that it produces different predictions for the same dataset. This is called bagging. The final desirable model is the average of boosting and bagging so that it can both predict as well as interpret at the same time.

In order to remove all the complexities, machine learning is combined with some techniques of deep learning, which lead to advancement in technology. These are generally referred to as the practical approach or theory-free approach as it does not involve much of an explanation as they are mostly self-explanatory. In deep learning, all the algorithms are equipped in such a way that they represent a particular aspect of data and enhance the prediction of that part of data. This is how the data can be easily predicted by using Ml techniques and interpreted with a little help of deep learning techniques without any increase in the complexity. This allows easy interpretation of even low quality and unstructured data, which is present in huge data sets and makes the understanding easy for both analysts and nontechnical users. This integrated approach allows all the features to run together as system and perform a relevant analysis. Various ML algorithms applied by distinguished researchers to solve various risks is explained in Table 3.1.

3.4.1 Challenges of Machine Learning in Risk Management

Apart from all the use cases and advantages of using ML in risk analysis and management, there exist some of the major issues that need to be taken into

TABLE 3.1

ML Algorithms for Risk Management

Risk Type	ML Algorithms
Credit risk	Neural Networks, SVM, KNN, Random Forest, Lasso regression, Cluster analysis
Liquidity risk	SVM, ANN, Bayesian Networks
Market risk	GELM, Cluster analysis, SOM, Gaussian Mixtures, cluster analysis
Operational risk	Nonlinear clustering method, Neural Networks, k-Nearest Neighbor, Naïve Bayesian, Decision Tree.
RegTech risk	SVM

consideration. Some of the challenges of machine learning in risk management that the industry faces are as follows:

(i) *Availability of data:* One of the most significant challenges is to obtain relevant as well as suitable data for processing. There is a lot of potential for machine learning packages for performing multiple tasks such as processing images and natural language by reading all types of data, but it is becoming difficult to maintain discipline in the data internally. Moreover, the sharing of data is tough as the information is stored in different systems with different restrictions that make the information retrieval even more challenging. Also, perhaps the information may not be stored formally.

(ii) *Availability of skilled staff:* With the evolution of new techniques, the requirement of skilled labor is quite demanding. The implementation, understanding, and working of advanced features and solutions is of high importance. Since providing training to the unskilled staff is time-consuming, there are some attempts made to solve this problem by offering a course to the aspirants by building a campus with around 7,000 people in India where there is frequent presence of these skills.

(iii) *Accuracy of machine learning techniques:* It is not as simple as just applying various machine learning techniques to manage the potential risk; rather, it requires continuous evaluation of the process in order to obtain optimal current solution. In order to ensure feasibility, multiple testing of the technique is required.

(iv) *Transparency and ethics:* This is one of the popular issues that usually occur. It is evident from the fact that the models work between the layers and remain hidden in the initial input and final output. This can lead to compliance problems while validating the model. It may even turn into the loss of a firm or convergence of the uniform ideal for trading, resulting in a lot of risk. There are many ethical issues as well such as unequal lending decision due to gender, race, and sexual discrimination.

3.4.2   Machine Learning Use Cases in the Financial Sector

This section analyzes and gives a detailed description of various use cases of machine learning in the field of risk management in the financial sector such as fraud detection, credit risk, supervision of conduct, and market abuse in trading (Tanwar, 2020).

(i) *Fraud detection:* Machine learning is now often used in the field of fraud detection, mainly to detect credit card fraud in the banking system. It has resulted in higher significant success rate till now. Banks have set up various monitoring and surveillance systems in order to ensure security. These systems keep a track on the payment fraud activities that take place more often. The fraud model engine works on the traditional historical payment algorithm that blocks the fraudulent trade as soon as it detects it.

   The training, testing, and authorization of machine learning algorithms become possible because of large data sets provided by the credit card payments. Moreover, various classification algorithms are trained with the help of historical data with identified fraud and nonfraud tags. This training is done with the help of large historical nonfraudulent data. The above historical payment data sets generate a clear view of features of the card by distinguishing them on the basis of transactions, owner, and history of the payment.

(ii) *Credit risk:* The term "credit risk" signifies the prospects of loss because of unsuccessful payments made by the lender. Therefore, the credit risk management system (CRMS) is implemented in order to meet the losses by analyzing the fairness of the bank and its capital along with the loan loss section at any instant of time. This system has enhanced the transparency as demanded by the financial crises. Transparency is carried by paying more emphasis on the regular examination of the knowledge of banks for its customers and credit risks.

   Suitable credit risk management improves the entire performance of the system and provides a competitive benefit. Various prediction models are made in order to make predictions regarding the kind of lender. This is where the machine learning techniques are implemented. These machine learning algorithms provide positive results and greater success rates in solving the problems including credit risks as well.

(iii) *Supervision of conduct and market abuse in trading:* Another use case of machine learning in the risk management system is the surveillance of the performance loopholes generated mainly by the traders working in the financial institutions. The various trading illegalities often lead to economic as well as status failure. In order to overcome such crimes and shortcomings, numerous self-operating systems have been developed in order to provide a check on the trading behaviors of dealers with increased accuracy and distinct ways to identify them.

Earlier, in the initial stages, the system's performance was limited to checking the behavior of a single trade. But with time, various new machine learning approaches are applied and the machines can now identify huge and complicated data and analyze the entire portfolios. Apart from performing this analysis, the systems are also able to link distinct trading facts via email flux, calendar items, developing check-in and check-out times, and calls as well. The automatic analysis is performed using deep learning and text mining techniques, which make the system machine readable.

3.5 Case Studies to Understand the Role of Machine Learning in Risk Management

i) Case Study 1: Banks

On average banks are prone to various risks such as credit risk, market risk, foreign exchange risk, foreign risk, frauds, etc. These risks are treacherous for the reputation of the banks and their worthwhile procedures. These risks are dormant privation to the banks and they can be questioned for the obligations on them. Risks are unpredictable and can develop irrespective of time. How well would it be if the risk can be specified or predicted? To specify the upcoming risks in advance, machine learning can be implemented. All the possible risks are first analyzed and reviewed. During the process of analysis of these risks a large amount of data is collected which is basically unstructured in nature. This data is collected from market information, customer reviews, metadata etc.

Machine learning is all about working with data. It processes this data for working machines. The machines will produce a desired output on the basis of the input processed. When machine learning along with some algorithms will be implemented on the large amount of unstructured data. The machine will interact on the basis of the data and the output will be generated. This generated output can be a sort of prediction of the upcoming risk that might rise up in the near future.

For the accurate results from the system, it is important the training of the large amount of unstructured data that is collected is done in a perfect way. After the training is finished the data is converted into structured data or the labeled data. On this final set of data various algorithms are processed. These algorithms basically mould the data into a framework of a process to take place on the basis of which the final output is generated. The type of data collected in the final output will also be of the same category.

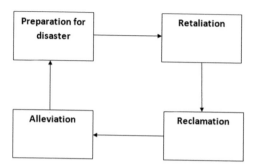

FIGURE 3.4
Role of machine learning in risk management for disaster management.

Machine learning is compatible, dynamic and can prevent a bank from such cunning risks. A number of problems can be easily fathom with the help of machine learning in the fields of risk assessment.

ii) Case Study 2: Disaster management

Disaster can be defined as the natural affliction that can lead to various human, environmental and property losses. These disasters are unpredictable but there are several technologies which are operated today to predict an approximate time for such calamities. The labeled data which is collected for machine learning is used to predict the disaster in the form of a sequence of their occurrence. There are various processes which can forecast the occurrence of the disasters on the basis of their probability of occurrence as shown in Figure 3.4. During these processes there is a chance that several fallacies occur. This measure indicates misfiring of appliance or apparatus, preservation excess, quantity of supplementary worked, etc.

These machines that predict the upcoming disasters and the human generation cannot afford risks in these indicators as everyone relies on their final output. Consequently, the role of risk assessment in fields like disaster management becomes necessary for protection of a specific area from sudden tragedy and irreplaceable losses. Machine learning includes several techniques that look after the prevention of such risks in this area.

Apart from all the issues and problems, looking at the positives in a situation is a much more crucial segment. Considering the future of machine learning in risk management is worthy. This is evident from the fact that the cost of development is reduced. Moreover, the time-consuming nature of this process has also decreased at a higher scale. One such instance is that of Banco Bilbao Viscera Argentaria (BBVA), a financial service company in Spain. This multinational corporation is investing heavily in compliance-based workers. Therefore, various machine learning techniques are used in order to reduce this outlay. This technology makes the repetitive tasks automatic. Along with the organization, data clustering as well as recovering of advanced data are some of the abilities of machine learning that will provide a huge benefit to the companies.

3.6 Conclusion

Throughout the chapter, various aspects and techniques of machine learning that are engaged in the fields of risk assessment to ameliorate this process has been discussed. The importance of machine learning in risk assessment is skillfully acknowledged. Machine learning has the potential to create wonders in risk assessment by incorporating methodical procedures and models that can fabricate accurate results for risk analysis by effectively monitoring large and complex datasets. The application of machine learning in this sector has led to the conclusion that these methods or techniques can be used to analyze huge amounts of data with efficient predictive analysis. Various use cases have also been discussed such as supervision of conduct and market abuse in trading, fraud detection, and credit risk.

The description of various models and the case studies that are covered in the chapter is a blueprint of the utilization of machine learning in different industries and organizations in the fields like risk management in banks and risk management in disaster management. The major issue that can be addressed by employing technology is to counter risks as machine learning provides various models and techniques that can minimize or prevent such risks. These techniques and models change with different sets of labeled data as per the convenience of the procedure. It provides the stability in the field and aids in the protection of human and national information that can be considered as highly confidential information. Still research is ongoing in the area of risk assessment and several measures are already implemented. There are still many areas untouched by machine learning. It can ease the stress of several industries and organizations as potential risks can be identified and suitable methods can be employed if the probability of risk is strong.

In the end, one can just say that humans will be encountering an era which will make even complex problems easy to solve with efficiency and they will solve it for sure, which will be time-saving and will be economically beneficial. Time is not far afar when technology such as AI and ML is employed as a solution to almost every problem.

References

Apostolakis, G.E. 2004. How useful is quantitative risk assessment? *Risk Analysis* 24: 515–520.

Aggarwal, P.K., Grover, P.S., Ahuja, L. 2018. Exploring quality aspects of smart mobile phone applications. *Journal of Advanced Research in Dynamical and Control Systems* 11: 292–297.

Aggarwal, P.K., Grover, P.S., Ahuja, L. 2019. Assessing quality of mobile applications based on a hybrid MCDM approach. *International Journal of Open Source Software and Processes*, 10: 51–65.

Aven, T. 2012. The risk concept—historical and recent development trends. *Reliab. Eng. Syst. Saf.* 99: 33–44.

Aven, T., Krohn, B.S. 2014. A new perspective on how to understand, assess and manage risk and the unforeseen. *Reliab. Eng. Syst. Saf.* 121:1–10.

Bucelli, M., Paltrinieri, N., Landucci, G., 2018. Integrated risk assessment for oil and gas installations in sensitive areas. *Ocean Eng.* 150: 377–390.

Chen, H., Moan, T., Verhoeven, H. 2008. Safety of dynamic positioning operations on mobile offshore drilling units. *Reliab. Eng. Syst. Saf.* 93: 1072–1090.

Cheng, H.-T., Koc, L., Harmsen, J., Shaked, T., Chandra, T., Aradhye, H., Anderson, G., Corrado, G., Chai, W., Ispir, M. 2016. Wide and deep learning for recommender systems. In: Proceedings of the 1st Workshop on Deep Learning for Recommender Systems. ACM, pp. 7–10.

Comfort, L.K. 2019. *The Dynamics of Risk: Changing Technologies, Complex Systems, and Collective Action in Seismic Policy*. Princeton, NJ: Princeton University Press.

Creedy, G.D. 2011. Quantitative risk assessment: how realistic are those frequency assumptions? *J. Loss Prev. Process Ind.* 24: 203–207.

De Marchi, B., Ravetz, J.R. 1999. Risk management and governance: a post-normal science approach. *Futures* 31: 743–757.

Durga Rao, K., Gopika, V., Sanyasi Rao, V.V.S., Kushwaha, H.S., Verma, A.K., Srividya, A. 2009. Dynamic fault tree analysis using Monte Carlo simulation in probabilistic safety assessment. *Reliab. Eng. Syst. Saf.* 94: 872–883.

Diekmann, E.J. 1992. Risk analysis: lessons from artificial intelligence. *Int. J. Proj. Manag.* 10, 75–80.

Goodfellow, I.J., Bengio, Y., Courville, A. 2016. *Deep Learning*. The MIT Press, Cambridge, MA.

Hastie, T., Tibshirani, R., Friedman, J. 2009. Unsupervised learning. In: *The Elements of Statistical Learning*. Springer, Berlin, pp. 485–585.

Haugen, S., Vinnem, J.E. 2015. Perspectives on risk and the unforeseen. *Reliab. Eng. Syst. Saf.* 137: 1–5.

Jain, P., Sharma, A., Ahuja, L. 2018. The impact of agile software development process on the quality of software product. In: Proceedings of the International Conference on Reliability, Infocom Technologies and Optimization (Trends and Future Directions) (ICRITO), pp. 812–815.

Jain, P., Sharma, S. 2019. Prioritizing factors used in designing of test cases: An ISM-MICMAC based analysis. In: Proceedings of International Conference on Issues and Challenges in Intelligent Computing Techniques (ICICT), India.

Jain, P., Sharma, A., Aggarwal, P.K. 2020. Key attributes for a quality mobile application. In Proceedings of the International Conference on Cloud Computing, Data Science and Engineering Confluence, 50–54.

Kaplan, S., Garrick, B.J., 1981. On the quantitative definition of risk. *Risk Anal.* 1: 11–27.

Khakzad, N. 2015. Application of dynamic Bayesian network to risk analysis of domino effects in chemical infrastructures. *Reliab. Eng. Syst. Saf.* 138: 263–272.

Khakzad, N., Khan, F., Amyotte, P. 2013a. Risk-based design of process systems using discrete-time Bayesian networks. *Reliab. Eng. Syst. Saf.* 109: 5–17.

Khakzad, N., Khan, F., Amyotte, P. 2013b. Quantitative risk analysis of offshore drilling operations: a Bayesian approach. *Saf. Sci.* 57: 108–117.

King, G., Zeng, L. 2001. Logistic regression in rare events data. *Polit. Anal.* 9: 137–163.

Kongsvik, T., Almklov, P., Haavik, T., Haugen, S., Vinnem, J.E., Schiefloe, P.M. 2015. Decisions and decision support for major accident prevention in the process industries. *J. Loss Prev. Process Ind.* 35: 85–94.

Landucci, G., Paltrinieri, N. 2016a. A methodology for frequency tailorization dedicated to the Oil & Gas sector. *Process Saf. Environ. Prot.* 104: 123–141.

Landucci, G., Paltrinieri, N. 2016b. Dynamic evaluation of risk: from safety indicators to proactive techniques. *Chem. Eng. Trans.* 53: 169–174.

Lasi, H., Fettke, P., Kemper, H.-G., Feld, T., Hoffmann, M. 2014. Industry 4.0. *Bus. Inf. Syst. Eng.* 6: 239–242.

Musgrave, G.L. 2013. James Owen Weatherall, the physics of Wall Street: A brief history of predicting the unpredictable. *Bus. Econ.* 48: 203–204.

Nivolianitou, Z.S., Leopoulos, V.N., Konstantinidou, M. 2004. Comparison of techniques for accident scenario analysis in hazardous systems. *J. Loss Prev. Process Ind.* 17: 467–475.

Nobre, F.S., 2009. *Designing Future Information Management Systems.* IGI Global.

Noh, Y., Chang, K., Seo, Y., Chang, D. 2014. Risk-based determination of design pressure of LNG fuel storage tanks based on dynamic process simulation combined with Monte Carlo method. *Reliab. Eng. Syst. Saf.* 129: 76–82.

Nývlt, O., Haugen, S., Ferkl, L. 2015. Complex accident scenarios modelled and analysed by stochastic Petri nets. *Reliab. Eng. Syst. Saf.* 142: 539–555.

Nývlt, O., Rausand, M. 2012. Dependencies in event trees analyzed by Petri nets. *Reliab. Eng. Syst. Saf.* 104: 45–57.

Øien, K., Utne, I.B., Herrera, I.A. 2011. Building safety indicators: Part 1 – theoretical foundation. *Safety Sci.* 49: 148–161.

Paltrinieri, N., Khan, F., Cozzani, V. 2015. Coupling of advanced techniques for dynamic risk management. *J. Risk Res.* 18: 910–930.

Paltrinieri, N., Reniers, G. 2017. Dynamic risk analysis for Seveso sites. *J. Loss Prev Process Ind.* 49: Part A: 111–119.

Paltrinieria, N., Comfortb, L., Reniers, G. 2019. Learning about risk: Machine learning for risk assessment. *Safety Sci.* 118: 475–486.

Pasman, H., Reniers, G. 2014. Past, present and future of quantitative risk assessment (QRA) and the incentive it obtained from land-use planning (LUP). *J. Loss Prev. Process Ind.* 28: 2–9.

Svozil, D., Kvasnicka, V., Pospichal, J. 1997. Introduction to multi-layer feed-forward neural networks. *Chemom. Intell. Lab. Syst.* 39: 43–62.

Tanwar, S., Bhatia, Q., Patel, P., Kumari, A., Singh, P.K., Hong, W.C. 2020. Machine learning adoption in blockchain-based smart applications: The challenges, and a way forward. *IEEE Access*. 8: 474–488.

Villa, V., Paltrinieri, N., Khan, F., Cozzani, V. 2016a. Towards dynamic risk analysis: a review of the risk assessment approach and its limitations in the chemical process industry. *Saf. Sci*. 89. https://doi.org/10.1016/j.ssci.2016.06.002

Villa, V., Paltrinieri, N., Khan, F., Cozzani, V., 2016b. A short overview of risk analysis background and recent developments. In: Dynamic Risk Analysis in the Chemical and Petroleum Industry: Evolution and Interaction with Parallel Disciplines in the Perspective of Industrial Application. 10.1016/B978-0-12-803765-2.00001-9.

Wolpert, D.H., 2002. The supervised learning no-free-lunch theorems. *Soft Comput. Ind.* 25–42.

Yang, X., Haugen, S., 2015. Classification of risk to support decision-making in hazardous processes. *Safety Sci.* 80: 115–126.

4

Machine Learning Techniques Enabled Electric Vehicle

Shyamalagowri Murugesan[1] and Revathy Jayabaskar[2]

[1]*Associate Professor, Department of Computer Science Engineering, Erode Sengunthar Engineering College (Autonomous), Erode, India*

[2] *School of Computing, SASTRA Deemed University, Thanjavur, India*

E-mail: mshyamalagowri2011@gmail.com; revathyjayabaskar@gmial.com

CONTENTS

DOI: 10.1201/9781003104858-4

4.1 Introduction

Machine learning is expected to play a vital role in the upcoming industry revolution. The evolution of ML and AI has great implications for the development of electric vehicles in various means. Battery performance can make or break the electric vehicle experience, from driving range to charging time to the lifetime of the car. Now, artificial intelligence has made dreams like recharging an EV in the time it takes to stop at a gas station a more likely reality and could help improve other aspects of battery technology.

For decades, advances in electric vehicle batteries have been limited by a major bottleneck: evaluation times. At every stage of the battery development process, new technologies must be tested for months or even years to determine how long they will last. But now, a team led by Stanford professors Stefano Ermon and William Chueh has developed a machine learning-based method that slashes these testing times by 98 percent. Although the group tested their method on battery charge speed, they said it can be applied to numerous other parts of the battery development pipeline and even to nonenergy technologies.

"In battery testing, you have to try a massive number of things, because the performance you get will vary drastically," said Ermon, an assistant professor of computer science. "With AI, we're able to quickly identify the most promising approaches and cut out a lot of unnecessary experiments."

4.1.1 Artificial Intelligence Technology to Enhance EV Production and Support the Deployment of Electric Vehicles

Electric vehicles reduce tailpipe emissions from the transportation sector and produce important public health benefits. One of the most important complementary assets for EV adoption is charging station infrastructure.

While we initially thought that the availability of real-time transactions from charging stations would be a primary innovation for this research, we quickly realized that we also had an enormous quantity of unstructured text to learn about how EV users engage with each other. Surprisingly until this point, the best evidence on EV consumer behavior typically relied on surveys or simulated data, and these were limited to a single market or region. We understood that data from mobile platforms offered a unique ability to aggregate large-scale evidence in ways not previously possible.

What was most exciting for our research team was the fact that the data was streaming and offered the potential to update in near real-time. This is in stark contrast to data collection from large government surveys, which take several years to complete, and data discovery is often slow and expensive. So, we set off to see if we could analyze both government-run and privately run EV charging stations to provide national, evidence-based analysis.

Given the quantity of data however, a major challenge for the team was the fact that it would take human experts about 32 weeks to sift through EV user reviews in order to extract useful insights from free text.

So, almost on the side, we started experimenting with deep learning natural language processing techniques to unlock some insights there. It turned out that the review data was an untapped source of research innovation for us. We quickly realized that the application of AI/ML to this data could both accelerate policy analysis and reduce science and technology (S&T) research evaluation costs.

Consequently, by deploying deep learning techniques to analyze those EV user reviews, we were able to show how machine learning tools could be used to quickly analyze streaming data for policy evaluation in near real time and provide new insight into important electric vehicle charging infrastructure policies, such as the need to focus on the quality of the user experience and evidence supporting public involvement in EV charging network buildout.

By displacing gasoline and diesel fuels, electric cars and fleets reduce emissions from the transportation sector, thus offering important public health benefits. However, public confidence in the reliability of charging infrastructure remains a fundamental barrier to adoption. Using large-scale social data and machine learning from 12,720 electric vehicle (EV) charging stations, we provide national evidence on how well the existing charging infrastructure is serving the needs of the rapidly expanding population of EV drivers in 651 core-based statistical areas in the United States. We deploy supervised machine learning algorithms to automatically classify unstructured text reviews generated by EV users. Extracting behavioral insights at a population scale has been challenging given that streaming data can be costly to classify. Using computational approaches, we reduce processing times for research evaluation from weeks of human processing to just minutes of computation. Contrary to theoretical predictions, we find that stations at private charging locations do not outperform public charging locations provided by the government. Overall, nearly half of the drivers who use mobility applications have faced negative experiences at EV charging stations in the early growth years of public charging infrastructure, a problem that needs to be fixed as the market for electrified and sustainable transportation expands.

The study, published by *Nature* on February 19, 2020, was part of a larger collaboration among scientists from Stanford, Massachusetts Institute of Technology (MIT), and the Toyota Research Institute that bridges foundational academic research and real-world industry applications. The goal: finding the best method for charging an EV battery in 10 minutes that maximizes the battery's overall lifetime. The researchers wrote a program that, based on only a few charging cycles, predicted how batteries would respond to different charging approaches. The software also decided in real time what charging approaches to focus on or ignore. By reducing both the length and number of trials, the researchers cut the testing process from almost two years to 16 days.

"We figured out how to greatly accelerate the testing process for extreme fast charging," said Peter Attia, who co-led the study while he was a graduate student. "What's really exciting, though, is the method. We can apply this approach to many other problems that, right now, are holding back battery development for months or years."

4.1.2 Artificial Intelligence Used to Supercharge Battery Development for Electric Vehicles

Battery performance can make or break the electric vehicle experience, from driving range to charging time to the lifetime of the car. Now, artificial intelligence has made dreams like recharging an EV in the time it takes to stop at a gas station a more likely reality and could help improve other aspects of battery technology.

For decades, advances in electric vehicle batteries have been limited by a major bottleneck: evaluation times. At every stage of the battery development process, new technologies must be tested for months or even years to determine how long they will last. But now, a team led by Stanford professors Stefano Ermon and William Chueh has developed a machine learning-based method that slashes these testing times by 98 percent. Although the group tested their method on battery charge speed, they said it can be applied to numerous other parts of the battery development pipeline and even to nonenergy technologies.

"In battery testing, you have to try a massive number of things, because the performance you get will vary drastically," said Ermon, an assistant professor of computer science. "With AI, we're able to quickly identify the most promising approaches and cut out a lot of unnecessary experiments."

4.1.3 A Smarter Approach to Battery Testing

Designing ultrafast-charging batteries is a major challenge, mainly because it is difficult to make them last. The intensity of the faster charge puts greater strain on the battery, which often causes it to fail early. To prevent this damage to the battery pack, a component that accounts for a large chunk of an electric car's total cost, battery engineers must test an exhaustive series of charging methods to find the ones that work best.

The new research sought to optimize this process. At the outset, the team saw that fast-charging optimization amounted for many trial-and-error tests—something that is inefficient for humans, but the perfect problem for a machine.

"Machine learning is trial-and-error, but in a smarter way," said Aditya Grover, a graduate student in computer science who co-led the study. "Computers are far better than us at figuring out when to explore—try new and different approaches—and when to exploit, or zero in, on the most promising ones."

The team used this power to their advantage in two key ways. First, they used it to reduce the time per cycling experiment. In a previous study, the researchers found that instead of charging and recharging every battery until it failed—the usual way of testing a battery's lifetime—they could predict how long a battery would last after only its first 100 charging cycles. This is because the machine learning system, after being trained on a few batteries cycled to failure, could find patterns in the early data that presaged how long a battery would last.

Second, machine learning reduced the number of methods they had to test. Instead of testing every possible charging method equally, or relying on intuition, the computer learned from its experiences to quickly find the best protocols to test.

By testing fewer methods for fewer cycles, the study's authors quickly found an optimal ultra-fast-charging protocol for their battery. In addition to dramatically speeding up the testing process, the computer's solution was also better—and much more unusual—than what a battery scientist would likely have devised, said Ermon.

"It gave us this surprisingly simple charging protocol—something we didn't expect," Ermon said. Instead of charging at the highest current at the beginning of the charge, the algorithm's solution uses the highest current in the middle of the charge. "That's the difference between a human and a machine: The machine is not biased by human intuition, which is powerful but sometimes misleading."

4.1.4   Wider Applications

The researchers said their approach could accelerate nearly every piece of the battery development pipeline: from designing the chemistry of a battery to determining its size and shape, to finding better systems for manufacturing and storage. This would have broad implications not only for electric vehicles but for other types of energy storage, a key requirement for making the switch to wind and solar power on a global scale.

"This is a new way of doing battery development," said Patrick Herring, coauthor of the study and a scientist at the Toyota Research Institute. "Having data that you can share among a large number of people in academia and industry, and that is automatically analyzed, enables much faster innovation."

The study's machine learning and data collection system will be made available for future battery scientists to freely use, Herring added. By using this system to optimize other parts of the process with machine learning, battery development—and the arrival of newer, better technologies—could accelerate by an order of magnitude or more, he said.

The potential of the study's method extends even beyond the world of batteries, Ermon said. Other big data testing problems, from drug development to optimizing the performance of X-rays and lasers, could also be revolutionized by the use of machine learning optimization. And ultimately, he said, it could even help to optimize one of the most fundamental processes of all.

"The bigger hope is to help the process of scientific discovery itself," Ermon said. "We're asking: Can we design these methods to come up with hypotheses automatically? Can they help us extract knowledge that humans could not? As we get better and better algorithms, we hope the whole scientific discovery process may drastically speed up."

4.1.5 A Review on AI-based Predictive Battery Management System for E-Mobility

With the advancement in digitalization and availability of reliable sources of information that provide credible data, Artificial Intelligence (AI) has emerged to solve complex computational real-life problems, which were challenging earlier. The artificial neural networks (ANNs) play a very effective role in digital signal processing. However, ANNs need rigorous main processors and high memory bandwidth, and hence cannot provide expected levels of performance. As a result, hardware accelerators such as graphic processing units (GPUs), field programmable gate arrays (FPGAs), and application-specific integrated circuits (ASICs) have been used for improving overall performance of AI-based applications. FPGAs are widely used for AI implementation as FPGAs have features like high-speed acceleration and low power consumption, which cannot be done using central processors and GPUs. FPGAs are also a reprogrammable unlike central processors, GPU, and ASIC. In electric-powered vehicles (E-Mobility), battery management systems (BMS) perform different operations for better use of energy stored in lithium-ion batteries (LiBs). The LiBs are a nonlinear electrochemical system, which is very complex and time variant in nature. Because of this nature, estimation of states like state of charge (SoC), state of health (SoH), and remaining useful life (RUL) is very difficult. This has motivated researchers to design and develop different algorithms that will address the challenges of LiB state estimations. This chapter intends to review AI-based data-driven approaches and hardware accelerators to predict the SoC, SoH, and RUL of the LiBs. The goal is to choose an appropriate algorithm to develop an advanced AI-based BMS that can precisely indicate the LiB states, which will be useful in e-mobility. An overview of the state-of-the-art on intelligent battery management systems for electric and hybrid electric vehicles is also provided. The focus is on mathematical principles, methods, and practical implementations. The intelligent battery management systems aim at lengthening the lifetime of the battery pack and enhancing the safety of drivers of electric and hybrid electric vehicles. Three major research topics are covered in the chapter, state of charge (SoC), state of health (SoH) of the battery pack, and the remaining driving range estimation.

4.2   Reverse Engineering with AI for Electric Power Steering

A prominent mystery of automobile industry scenario is that the vehicle creators are more curious and covetous of knowing each and every minuscule insight concerning their rival's vehicles. To discover the subtleties generally includes the study of start of art of the product and the concept of dismantling a whole vehicle, piece by piece. The vehicle creators either have their own inside groups who purchase a contender's vehicle and dismantle it or pay an external organization to do likewise. The reverse engineering of vehicles is a concept that is carried out by some companies who well specialized in it. They take pleasure in purchasing the most recent model of any vehicle and fastidiously dismantling it like so numerous Lego blocks.

Apart from the delight, the reverse engineering concept can put forth some great bucks by their attempts. They will offer solutions concerning their rival's vehicles to other vehicle creators. A full reported document of each thing that went into the vehicle, alongside valuable added perspectives like how much every segment probably costs. Moreover, they can assess the amount it costs to accumulate the parts and give car producer knowledge into what sort of gathering exertion their opposition is utilizing.

Manufacturers need to pay more attention to get entire knowledge about their own vehicles and would definitely know how they accumulate their own vehicles and what it costs. However, it tends to be extremely convenient to see the evaluations made by the outsider. The customer may be high or low, or in any case not all that demanding. It is additionally helpful to understand what they are telling the rivals. In addition, it would then be able to have them do a correlation for the parts and expenses related with the rivals versus the vehicles.

A reverse engineering could convey whatever the client needed to think about the vehicle, even the weight, size, and cost of each part along with the ability of that part. The data about the producer of the part and piece of the part is, for example, rates of metal versus different components could also be given. Everyone discovers overpaying for their own carburetor and realizes that to reduce costs and ultimately reduce the price, they charge for their car that ought to switch. There's an additional contort to this figuring out exertion. In principle, in the event of being cautious to dismantle a vehicle, likewise examine the parts into a CAD framework and basically figuring out the whole plan of the vehicle. With the present sophisticated CAD frameworks, it would then be able to see the vehicle from any point and jump into and zoom all through the entire plan of the vehicle. Some truly modern CAD frameworks will permit to siphon the plan into a test system program. This could permit to conceivably go about like the vehicle exists and perceive how it runs.

At the Cybernetic AI Self-Driving Car Institute, they are creating AI frameworks for self-driving vehicles, which reveal an unmistakable fascination that a similar industry-wide quest for figuring out of customary vehicles is presently in progress for AI self-driving vehicles as well. The five key stages to the preparing parts of an AI self-driving vehicle are sensor information assortment and translation, sensor combination, virtual world model refreshing, AI activity plan refreshing, and car orders controls issuance. The most noticeable actual parts of an AI self-driving vehicle are the sensors. There are a huge number of sensors on an AI self-driving vehicle, including radar sensors, sonic sensors, cameras, light detection and ranging (LIDAR), inertial measurement units (IMUs), and different sensors. Every car creator and tech firm is definitely keen on knowing which sensors different organizations are utilizing in their AI self-driving vehicles. It's a fairly free-for-good now in that there is no predominant provider fundamentally. To be sure, the different organizations that make sensors appropriate for AI self-driving vehicles are in a savage battle about attempting to pick up gain traction for their sensors.

4.3 Artificial Intelligence Technology to Enhance EV Production and Support the Deployment of Electric Vehicles

The new AI innovation, which has been made via Artificial Intelligence, is intended for modern scale EV assembling to improve the development and assembling of batteries. EV battery packs are made out of an enormous number of individual battery cells that are held and electronically associated by various welded joints. High electrical resistance, because of helpless joint quality, can make energy misfortune and warmth age, which can cause a potential well-being issue and furthermore lessens the productivity of the battery. The AI innovation consequently evaluates the surface imperfections and the electrical resistance of each joint before conclusive get together, accordingly saving the maker's time and expenses while additionally guaranteeing the well-being of the battery.

Along with the improvement of smart grids, the wide appropriation of electric vehicles (EVs) is viewed as a motivation to the decrease of CO_2 outflows and more intelligent transportation systems. Specifically, EVs expand the grid with the capacity to store energy at certain point in the network and give it back at others and, accordingly, help to improve the utilization of energy from irregular environmentally friendly power sources and let clients refill their vehicles in an assortment of areas. Nonetheless, various challenges should be tended to if such advantages are to be accomplished. From one viewpoint, given their restricted reach and costs associated with charging EV batteries, it is essential to plan calculations and methodology that will limit costs and, simultaneously, avoid clients being standard. Then again, groups of EVs should be coordinated so as to stay away from peak on the grid that may bring about high electric tariff costs and overburden local distribution grids.

4.4 Artificial Intelligence Used to Supercharge Battery Development for Electric Vehicles

Battery operation can represent the deciding moment for the electric vehicle experience, from driving reach to charging time to the lifetime of the vehicle. Presently, artificial intelligence has made dreams in energizing an electric vehicles in the time it takes to stop at a gas station a more likely reality and could help to improve other aspects of battery technology. For quite a long time, propels in electric vehicle batteries have been restricted by a significant bottleneck—assessment times.

Designing ultrafast charging batteries is a significant test, essentially in light of the fact that it is hard to make them last. The intensity of the quicker charge puts more prominent strain on the battery, which regularly makes it fail early. To prevent this harm to the battery pack, a part that represents an enormous piece of an electric vehicle's all out cost, battery engineers should test a comprehensive arrangement of charging techniques to locate the ones that work best. The new exploration tried to upgrade this cycle. At the start, the group saw that quick charging enhancement added up to numerous experimentation tests, something that is inefficient for people, yet the ideal issue for a machine. Machine learning is experimentation, however in a more intelligent way and PCs are much better than us at sorting out when to investigate, attempt new and various methodologies and when to adventure, or focus in, on the most encouraging ones. The group utilized this capacity for their potential benefit in two key manners.

To start with, they utilized it to lessen the time per cycling test. In a past report, the scientists found that as opposed to charging and energizing each battery until it failed, the standard method of testing a battery's lifetime. They could predict how long a battery would last after just its initial 100 charging cycles. This is on the grounds that the AI framework, subsequent to being prepared on a couple of batteries cycled to disappointment, could discover designs in the early information that foretold how long a battery would last. Second, machine learning diminished the quantity of strategies they needed to test. Rather than testing each conceivable charging technique similarly, or depending on instinct, the PC gained from its encounters to rapidly locate the best conventions to test.

By testing less strategy for less cycle, the examination's creators immediately found an optimal ultrafast charging convention for their battery. Notwithstanding significantly accelerating the testing cycle, the PC's answer was likewise better and substantially more uncommon than what a battery researcher would almost certainly have conceived. It gave us this shockingly straightforward charging convention something we didn't expect. Rather than charging at the most elevated current toward the start of the charge, the calculation's answer utilizes the most noteworthy current in the charge. That is the distinction between a human and a machine: The machine isn't one-sided by human instinct, which is ground-breaking yet some of time deceiving.

The specialists said their methodology could quicken virtually every bit of the battery advancement pipeline: from planning the science of a battery to deciding its size and shape, to finding better frameworks for assembling and capacity. This would have wide ramifications for electric vehicles as well as for different sorts of energy stockpiling, a critical necessity for doing the change to wind and sunlight-based force on a worldwide scale. The investigation's AI and information assortment framework will be made accessible for future battery researchers to uninhibitedly utilize, Herring added. By utilizing this framework to improve different pieces of the cycle with AI, battery advancement and the appearance of fresher, better innovations could quicken by a significant degree or more. The capability of the investigation's strategy broadens even past the universe of batteries.

Advances in electric vehicle batteries have regularly been restricted by the colossal bottleneck of assessment times. At each phase of the battery advancement measure, new advances should be tried for quite a long time or up to years to decide how long they will last. Presently, Stanford University analysts have built up an AI-based strategy that cuts these testing times by 98 percent. The specialists tried the strategy on battery charge speed, however state that it tends to be applied to numerous different pieces of battery improvement. The advancement strategy could make the fantasies about energizing an electric battery in the time it takes to stop at a service station a more probable reality. Artificial Intelligence is ready to add some juice to electric vehicles by accelerating upgrades in battery innovation.

Specialists are drawing nearer to conveying batteries that highlight upgrades profoundly looked for by architects and advertisers of electric vehicles: batteries that are more secure, revive quicker, and are more practical than the age of lithium-ion batteries now being used. Within five years, specialists state, electric vehicles will arrive at value equality with conventional burning motor vehicles, thanks to some extent to the job AI is playing. The battery makes up 25 percent of an EV's complete expense, and half of that battery cost is materials. Utilizing AI, battery analysts can accelerate the investigation and testing of a tremendous universe of new material definitions that could make up a battery. The outcome is that progresses in innovative work are going on at a speed up till now incomprehensible. Disclosures that once would have taken a long time to accomplish now are conceivable, now and again, very quickly.

4.5　How AI Helps Build Better Batteries

Working on quicker charging, more secure, and all the more impressive electric vehicles, battery scientists are utilizing AI to accelerate the quest for better materials and the testing cycle. Beneath, a glance at how AI could improve lithium-ion batteries. To comprehend the progressions occurring, it assists with taking a gander at how ebb and flow battery research is directed, and at lithium-ion batteries, a battery type that packs a great deal of energy into a little bundle, and that controlled the unrest in purchaser gadgets and EVs.

Lithium-ion batteries have three fundamental segments, yet each is made out of numerous materials. This makes a practically boundless universe of conceivable substitution materials when researchers need to dabble to improve value, well-being, or execution. Prior to AI, a quest for new materials was finished with conventional measurable examination, which was restricted by processing capacity and a specialist's capacity to gather data from the information. Man-made intelligence, or AI, changes the entirety of that. Quick PCs prepared to perceive explicit examples are filtering through immense measures of information put away in the cloud, taking advantage of materials' information bases, analysis of results, and long periods of logical writing, in a powerful hunt to recognize which battery sciences are probably going to beat and which are probably going to come up short.

At its examination lab in Almaden, California, International Business Machines Corp. (IBM), has 1.94 percent researchers utilizing AI to quicken improvement of a lithium-ion battery that is quick charging and liberated from hefty metals. Instead of utilizing nickel and cobalt, which are costly, in restricted stock and hard to reuse, IBM says it has found an approach to utilize an iodine-based material that can be sourced from brackish waters found in saltwater, conceivably making this battery a lot less expensive to create than the present lithium-ion details. One late arrangement of potential materials for use in the electrolyte, a substance inside a battery that helps with the age of the electrical flow *and* had around 20,000 likely mixes. Utilizing customary computational techniques to screen those competitors might have taken five years. AI assessed them in nine days. The AI is currently getting substantially more important as we change the solvents and electrolytes to improve regarding limit and life cycle, since we have had more opportunity to prepare it.

Notwithstanding assisting with materials research, AI decreases the time it takes to test batteries. Previously, streamlining new battery-cell plans was a cycle that regularly took long stretches of charging and releasing the batteries a great many occasions. Presently, AI's capacity to rapidly investigate tremendous measures of information gathered during battery testing permits the researchers to make forecasts about execution a lot quicker, decreasing the quantity of tests that should be run. The anode stores lithium and deliveries lithium-ion (lithium less an electron) when the battery is releasing. The separator permits lithium ion to go through while electrons are compelled to travel independently delivering an electric flow. What's more, the AI empowers specialists to recognize incorporated natural mixes that could help the anode's ability to hold lithium ion. The cathode works when the battery is revived, and the lithium ion and electrons venture out back to the anode in different ways. What's more, with AI's assistance, scientists are building up an iodine-based option in contrast to cobalt, a costly, hard to reuse hefty metal utilized in most lithium-ion batteries. The electrolyte, normally a blend of salts and solvents, encourages the development of the lithium ion. With AI scientists can test new electrolyte plans quicker, assisting them with distinguishing mixes with expanded voltage and higher blaze focuses. Quicker charging commonly harms the separator, lessening the battery's lifetime and possibly prompting fires. Computer-based intelligence assists analysts with finding the sweet spot adjusting charging speeds, charging flows, charging recurrence, and battery lifetime.

Researchers and analysts around the globe are progressively going to Artificial Intelligence (AI) and alleged robo-scientists to assist them with finding everything from new anti-infection agents and medications, through to new shower on sun-powered board materials and immunizations, so it shouldn't come as a very remarkable amazement that AI is presently being utilized to help grow new battery tech. "In battery testing, you need to attempt countless things, on the grounds that the exhibition you get will change definitely," said Ermon, an associate educator of software engineering at MIT, alluding to how researchers customarily chase for new battery forward leaps, and who drove the new undertaking to utilize AI to assist them with creating promising batteries for electric vehicles. "With AI, we're ready to rapidly recognize the most encouraging methodologies and cut out a great deal of pointless trials." The examination, distributed by *Nature*, was essential for a bigger cooperation among researchers from Stanford University, MIT, and the Toyota Research Institute that spans basic scholastic exploration and genuine industry applications, and their objective: to locate the best strategy for charging an EV battery in a short time that amplifies the battery's general lifetime.

4.6 Machine Learning Supercharges Battery Development

The analysts composed an AI program that, based on few charging cycles, anticipated how batteries would react to various charging approaches. The product likewise chose continuously what charging ways to deal with center around or disregard. By lessening both the length and number of preliminaries, the analysts cut the testing cycle from very nearly two years to only 16 days.

4.7 AI-Based Predictive Battery Management System for E-Mobility

This chapter additionally features the specialized difficulties and arising innovations for the improvement of effectiveness, unwavering quality, and security of EVs in the coming stages as another commitment. The advantages of electric vehicles rise up out of these vehicles' capacity of continuing their energy requests through electric network as opposed to petroleum product utilization. And also environmental examinations have demonstrated that electric drive (E-drive) offers the most elevated eco-friendliness and subsequently the least discharge of greenhouse gases. The worldwide stores of diesel, petroleum, and other nonrenewable energy sources are diminishing quickly because of their broad use in transportation activity. The conventional power sources produce huge loads of CO_2 yearly, which have hurtful ramifications for the climate, for example, ozone-depleting substance discharges and a dangerous atmospheric deviation. Moreover, the expenses of these powers are expanding dramatically, so there is a requirement for an auxiliary fuel hotspot for transportation, for example, electric vehicles (EVs), new energy vehicles (NEVs), plug-in hybrid vehicle (PHEVs), battery electric vehicles (BEVs), and fuel cell electric vehicles (FCEVs). As of late, battery-powered batteries have pulled in significant consideration inferable from their appeal in EVs, HEVs, and PHEVs. With the utilization of sustainable power, these transportation sources can decrease their GHGE by up to 40 percent. The elective fuel sources, for example, wave, wind, flowing, and sunlight-based, are episodic, so these energy assets additionally require an energy storage systems (ESS) to keep a smooth and dependable inventory.

The battery management systems (BMS) is a basic part of electric and hybrid electric vehicles. The motivation behind the BMS is to ensure protected and dependable battery operation. To keep up the security and unwavering quality of the battery, state observing and assessment, charge control, and cell adjusting are functionalities that have been actualized in BMS. As an electrochemical item, a battery operates contrastingly under various operational and natural conditions. The vulnerability of a battery's exhibition represents a test to the execution of these capacities.

With the evolution in digitalization and accessibility of reliable sources of data that give sensible information, Artificial Intelligence (AI) has been developed to tackle a complex computational genuine issue, which was inspiring before. The ANNs assume an extremely compelling job in digital signal processing. ANNs need thorough principal processors and high memory data transfer capacity, and thus can't give anticipated degrees of execution. The graphic processing units (GPUs), field programmable gate arrays (FPGAs), and application-specific integrated circuits (ASICs) have been utilized for improving the overall performance of AI-based applications. FPGAs are generally utilized for AI usage as FPGAs have highlights like rapid quickening and low power utilization, which is impossible when utilizing local processors and GPUs. FPGAs are additionally a reprogrammable dissimilar to local processors, GPU, and ASIC. In electric-powered vehicles (e-mobility), BMS perform various tasks for better utilization of energy stored in lithium-ion batteries (LiBs).

The LiBs are a nonlinear electrochemical system, which is extremely complicated and time variant in nature. Due to this nature, assessment of states like state of charge (SoC), state of health (SoH), and remaining useful life (RUL) is challenging. This has inspired scientists to plan and create various calculations that will address the difficulties of LiB state assessments. Battery capacity should be intended to permit EV drivers arrive at their objective endpoint while evading unnecessary stops to energize their vehicles. Nonetheless, this extra battery capacity would affect the vehicle's space, weight, and cost. Considering these restrictions, the coordination of EVs with the vision of intelligent transportation systems (ITS) is integrated.

This section begins by placing the plan of EVs into a more extensive viewpoint by proposing a Predictive Intelligent Battery Management System (PIBMS), which will upgrade the general exhibition of EVs ,including energy utilization and outflows utilizing the ITS foundation. It spreads out the plan establishment for the future execution of an interconnected EV furnished with PIBMS, which further adds to the improvement of energy proficiency and diminished emanations. Energy stockpiling framework (ESS) innovation is as yet the logjam for the electric vehicle (EV) industry. Lithium-ion (Li-ion) batteries have pulled in impressive consideration in the EV business attributable to their high energy density, life expectancy, nominal voltage, power density, and cost. In EVs, a smart battery management system (BMS) is one of the fundamental parts; it gauges the conditions of battery precisely, yet in addition guarantees safe activity and drags out the battery life. The exact assessment of the condition of charge (SOC) of a Li-ion battery is an extremely testing task in light of the fact that the Li-ion battery is an exceptionally time variation, nonlinear, and complex electrochemical framework.

This section clarifies the functions of a Li-ion battery, gives the fundamental highlights of a smart BMS, and thoroughly surveys its SOC assessment techniques. These SOC assessment techniques have been ordered into four fundamental classes relying upon their inclination. A critical clarification, including their benefits, restrictions, and their assessment errors from different examinations, is given. A few proposals relying upon the advancement of innovation are recommended to improve the online assessment. This section tends to the worries for the current BMSs. State assessment of a battery, including condition of charge, condition of well-being, and condition of life, is a basic undertaking for a BMS. Through inspecting the most recent systems for the state assessment of batteries, the future difficulties for BMSs are introduced and potential arrangements are proposed too.

4.8 Uses of Artificial Intelligence for Electric Vehicle Control Applications

The electric vehicle (EV) is emerging as the best-in-class innovation vehicle tending to the consistently squeezing energy and environmental concerns. To diminish the dependence of oil and natural contamination, the development and integration of electric vehicles has been quickened in numerous nations. The execution of EVs, particularly battery electric vehicles, is viewed as an answer to the energy emergency and natural issues. This chapter provides significance to specialized improvement of EVs and arising advances for their future application. Key advances with respect to batteries, charging innovation, electric engines and control, and charging foundation of EVs have been summed up.

This section presents a novel speed control plan of electric vehicle (EV) to improve the comportment and strength under the condition of various road requirements. Parameters that direct the working of PI regulator are progressively changed with the help of fuzzy intelligence control. All things considered, electric vehicles (EVs), including full cell and hybrid vehicles, have been developed quickly as an answer for the energy and ecological issues. Driven EVs are controlled by electric engines through transmission and differential gears, while straightforwardly determined vehicles are pushed by in-wheel or, basically, wheel engines. The fundamental vehicle configurations of this exploration have two straightforwardly determined wheel engines introduced and worked inside the driving wheels on a pure EV. These wheel engines can be controlled autonomously and have so quick and exact reaction to the order that the vehicle skeleton control or movement control turns out to be more steady and vigorous, contrasted with, by implication, driven EVs. Like most exploration on the torque distribution control of the wheel engine, the wheel engines provided a powerful ideal tractive power conveyance control for an EV driven by four-wheel engines, in this manner improving vehicle handling and stability. The analysis has demonstrated that EV control techniques, for example, PI control can perform optimally over the full range of operations and disturbances and it is viable with consistent vehicle torque. Moreover, the nonlinear vehicle torque is not fixed and changes arbitrarily. Anyway, EV with regular PI control might not have agreeable execution in such quick fluctuating conditions, the system performance deteriorates. Also, it is hard to choose appropriate control boundaries K_p and K_i to accomplish acceptable compensation results while keeping up the stability of EV traction, because of the highly complex, nonlinear nature of controlled systems. These are two of the significant disadvantages of the PI control. To beat these challenges, adaptive PI regulator by fuzzy control has been applied both in fixed and under streets limitations and appears to improve the general presentation of EV.

The direct torque control (DTC) methodology is one sort of superior performance innovations for AC motors, because of its simple structure and capacity to accomplish quick reaction of flux and torque, due to which it has attracted interest in the recent years. DTC-SVM with PI regulator direct torque control without hysteresis band can successfully diminish the torque ripples; however, its system robustness will hide its improvement. DTC-SVM strategy can improve the system robustness, clearly diminish the torque and flux ripple, and adequately improve the dynamical responses. The DC–DC converter is used with a control methodology to guarantee the energy need for the EV and the drive system. The objective of this section is to add to understanding the intelligent fuzzy PI regulator for utility EV tow rear determining wheel applied direct torque control-based space vector balance under a few situations.

The commercialization of electric vehicles/hybrid electric vehicles (EV/HEV) is up to now being tested with respect to performance and cost. The Cost decline should be investigated as an ideal plan. Other featured research includes green drive plans, which incorporate the integration of possible supportable renewable energy sources to make eco-obliging green vehicles, similar to the Internet of Things (IoT)-based methods for EV/HEVs. Electric vehicle research incorporates multi-disciplinary capacity from electrical equipment.

4.9 Conclusion

There is an increasing demand for scalable and autonomous management systems. We have proposed AI-based EV, which would eliminate fossil fuel (petrol and diesel) consumption. A universal approach to performing the reverse engineering of electric power steering (EPS) for the purpose of external control is also presented, designed, and executed. The primary objective of the related study was to solve the problem associated with the precise prediction of the dynamic trajectory of an autonomous vehicle, which was presented and designed. This task has been accomplished by deriving a new equation for determining the lateral tire forces and adjusting some of the vehicle parameters under road test conditions. The expert systems were made more flexible and effective for the present application by the introduction of hybrid artificial intelligence with logical reasoning. The innovation offers a solution to the major problem of liability in the event of an autonomous transport vehicle being involved in a collision.

5

A Comparative Analysis of Established Techniques and Their Applications in the Field of Gesture Detection

Muskan Jindal, Eshan Bajal, and Shilpi Sharma
Department of Computer Science and Engineering, Amity University, Uttar Pradesh, India

E-mail: muskan.jindal1@student.amity.edu, eshan.bajal@student.amity.edu, ssharma22@amity.edu

CONTENTS

DOI: 10.1201/9781003104858-5

5.1 Introduction

In day-to-day mundane conversations and tasks, information reinforcement and clarification play a cardinal role, using sign language as a medium to do so. Sign language is an irreplaceable component of daily communication, and it may have variegated forms like visual motions or hand gestures, but it aims to convey the same message and fulfil the purpose of information reinforcement. It includes use of multiple body parts including fingers and hands, which are the popular ones, but it also incorporates facial gestures, head, body and arms, which may be lesser known in mundane communities. For those with hearing, listening or visual disabilities, sign language is the only means of communication and expression. But those with these disabilities depend completely on sign language to interact with outer world find it cumbersome to do so as many people are unaware of the complete imbrications of sign language. Moreover, while use and knowledge of sign language is common among communities with visual disabilities, it is not very widespread among communities with hearing disabilities. This poses a challenge for these communities to interact or convey their message to the outer world as they have relied completely on sign language, which continues to be an unsolved issue till today.

While most of the sign language includes upper half of body [1], some sign language methods use variegated shapes and multiple changes in the complete body [2]. Gestures most commonly used are hand gestures, which can be classified into multiple types, namely, conversational gestures, controlling gestures, manipulative gestures and communicative gestures [3]. Due to the highly structured and organized attributes of sign language, the field of computer vision and its various algorithms suits the same [4]. The study here compiles contemporary expertise on the detection of gestures implementing supervised learning and image segregation algorithms, along with their associated mannerisms while proposing modifications for inclusivity of a wider audience, with a prominent emphasis on people with communication disabilities. Furthermore, the work done in this chapter may culminate in the development of real-world applications such as determining the attentivity of students in real time, which is an indispensable asset in the post-COVID world of online interactions. Implementation of computer vision algorithm for gesture analysis is suitable for the complete process of identifying, monitoring and tracking various gestures performed while communicating in sign language because the complexity and structured nature of the process can be best interpreted by the nature of computer vision methodologies. Though the field seems inchoate and nascent, some of the earliest research literature was published in 1933, where multiple gesture analysis techniques were utilized to acclimate to available speech and handwriting material for further analysis and identification. Few of the same include technique implemented by Pentland and Darrell, which aided in the identification of multiple dynamic structures by using an optimized dynamic time warping (DTW). Eventually [5], hidden Markov models (HMMs) were implemented to comprehend and analyse various subtle attributes or properties to eventually categorise the shape and trajectory of various sign languages. Further, another literature elucidated the accuracy of the same by collating 262 signatures from variegated signers to obtain and calculate to an accuracy of around 94 percent, proving the method viable. The reason behind deliberating signs from multiple signers was that it was observed that considering signs from a particular individual though increased accuracy on training set but decreased the actual accuracy of model [6]. Later building on the concepts of HMMs, another pair of prolific researchers, namely, Vogler and Metaxas [7] elucidated the multiple limitations of HMMs while using the dependent model. [8] Ascension Technologies Flock of Birds devices collated three-dimensional translation and rotation data of the sign by the aid of multiple authors to facilitate researchers to further analyse the area of research.

Another very popular and well-developed area of gesture analysis, specifically hand gesture hand detection, is human–computer interaction, via a UI/UX based interface facilitated by a processing unit, where data is collected via visual or sensory devices like camera by video or imageries [9]. Multiple receptor devices like the one that can be worn for data acquisition include data gloves [10] that facilitate accurate and abundant access to data for gesture analysis or detection. Other sensory receptor devices, including camera, solve the issues like moment of hand gestures, which are the limitation of data gloves but create their own indigenous problems like camera orientation, background lights and foreground disturbance. Thus, data acquisition remains a cumbersome step in the gesture detection process but can be eased by multiple devices and frameworks like Microsoft Kinect [11], Leap motion controller [12] and ASUS Xtion [10]. The next step is to effectively learn the acquired data to comprehend and identify various patterns, similarities and deduce conclusions from the same. This task of learning from raw data can be performed easily by the application of variegated learning algorithms like deep learning, convolutional neural network (CNN), adapted convolutional neural network (ADCNN) and recurrent neural network (RNN) [13,14].

5.2   Challenges and Areas of Improvements

While there are multiple steps of gesture analysis as described by multiple authors, some gradation steps create more challenges than other methods. After discerning through multiple research articles and literature published till date, given below are the selected methods that create more challenges in the process of gesture analysis and recognition.

5.2.1   Image Acquisition Challenges

Capturing or acquiring images is done via multiple sources by the aid of camera or other image-capturing devices. This is the primary step of any gesture analysis or recognition framework and algorithm. When images are captured from an analysis perspective, for example those captured by a person or researcher, physical aspects like angle of picture and background colour contrast are taken care of, but when this task is performed by an automated machine, image capturing might not be done effectively [15]. Issues like image capturing frame and background colour contrast can create further issues in steps like gesture detection and unique gesture identification. While hand gesture detection frameworks have become robust and effective, computer vision is still an abstruse concept to people and image-capturing cameras are placed properly by the individual using the same. So, the *placement of camera* creates issues in later phases of gesture analysis, detection and identification [16]. Other issues in image capturing include loss of depth, varying speed of the gesture performer, inability to identify the difference between two gestures and spatial or temporal inconsistencies [17]. While the above-mentioned constraints are specific to a scenario, there are some very mundane issues that occur during the same including complicated or busy background, loss of focus due to busy backdrop, real-time monitoring constraints and others. The image capturing object, that is, camera and its specifications, also plays a vital role in image acquisition and its attributes such as colour, quality, resolution and number of frames per unit time or frame rate. 3D image capturing amplifies these constraints as more accurate and high-resolution camera is required here.

5.2.2   Gesture Tracking, Segmentation and Identification Challenge

Gestures have a certain degree of rotation, that is, the degree or directional radius where they move freely; the higher is their degree of rotation, higher is the complexity in gesture analysis and identification. While this appears as the major challenge especially in hand gestures, other limitations include inconsistent speed of gestures [18], inability to distinguish between start and end of gestures and variation on skin texture or complexion. Other gesture image segmentation challenges include inconsistencies on background like variation in illumination, busy or moving background, and continuous differences in colour patterns of the background [19].

5.2.3 Feature Extraction Challenges in Gesture Detection and Identification

There are multiple features that are extracted via variegated approaches that face varied challenges and graph-based features when extracted face limitations due to complex background or inconsistent illumination [19]. Mixing of gestures or mixing of features also creates several limitations in the same. While the degree of rotation is well accustomed with various gesture detection algorithms, abrupt gesture or out-of-defined boundary gestures create issues. This gradation of limitation also incorporates the challenges and issues that occur during the identification of continuous gestures as the detector deliberates them as one continuous gesture instead of multiple unique gestures. A solution to this issue of continuous gestures was proposed by introducing automatic system-based pauses since these pauses may or may not coincide with the actual pauses taken by the person. Moreover, many times due to personal habits, some people perform gestures by taking few or no pauses in between while others perform gestures with an optimal number of pauses [20]. These discrepancies also create inconsistences in the identification of gestures when gestures are performed by humans, but if performed by artificial devices, such issues of inconsistencies tend to be avoided [21]. Apart from the above, another type of gestures, namely 'conversational gestures', also have similar issues on grounds of gesture identification and segmentation as they have multiple variations based the personal way of using conversational gestures, variation in dialect as different dialects and regional influences create variations in speed, frequency and type of conversational gestures used. Real-time conversational gestures also differ widely based on the environment like public gestures and private gestures, which can create further discrepancies in both gesture identification and segmentation. Additionally, emotional influence on gestures is a factor that creates inconsistencies in the same [22,23]. There are categories of gestures that create problems of conscious and unconscious gestures that humans perform during mundane conversations. While this can be considered as a subcategory to conversational gestures, they can be deliberated more effectively when considered individually [24]. Thus, real-time gesture identification, analysis and segmentation is a cumbersome and complex task, which can be improved by continuous training over a long period of time.

5.2.4 Limitations and Challenges in Gesture Gradation or Categorization

A gesture-detecting algorithm's final gradation is to categorise the accumulated gestures into multiple categories, namely, supervised and unsupervised. Challenges occurring during this phase occur mainly during handling large data points like higher complexity data points, categorising ambiguous data into pre-defined classifications, identification of appropriate data sets that satisfy most of the data points, handling outlier data points and other limitations like handling ambiguous gestures [25]. Multiple challenges occur while implementing variegated classifiers and filters when classifying captured gestures. When implementing classifiers like dynamic time wrapping (DTW) classifier, paucity of huge number of data sets and vocabulary size of the same create a limitation. For distance-dependent algorithms like k-nearest neighbour, the large data set increases the computation complexity exponentially, thereby increasing time and space complexity as well, depleting the overall performance of the algorithm. Much like DTW algorithms, deep learning-based classification approaches also deplete the performance due to paucity of data sets and also require huge training or validation set to provide optimal performance. Primordial or classical methods of classification like finite-state machines have very rigid transform states and boundaries, which is a palpable limitation of the same. Cluster-based classifiers like k-means classification approach is predominantly dependent on the initial cluster and its attributes, which create discrepancies later, and also it is influenced negatively by the presence of classifiers. So choosing an appropriate value for 'k' is a cumbersome task. The similar issue of selecting optimal value is faced by sister classifiers known as support vector machine (SVM), where selecting appropriate kernel value becomes a daunting task. Moreover, SVM has other limitations like its high computation complexity, which increase the expenses, and being a binary classifier by nature, pairwise classification is required for handling multiple class classification as illustrated in Figure 5.1 below.

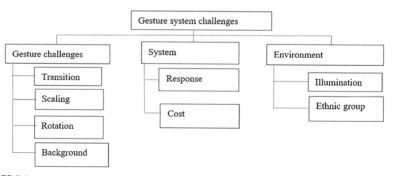

FIGURE 5.1

Limitations and Challenges in Gesture Gradation or Categorization are elaborated pictorially.

5.2.5 Limitations of End-User Gesture Analysis Customization

This is a comparatively special and new development in the field of gesture analysis, recognition and segmentation. It deliberates the scenario where the end-user requests for custom-specific gestures to be incorporated and further activities based on the same [26]. Wobbrock created a set of 27 novel gestures along with their respective activities performed as reaction to those gestures. Post that, many authors started adding custom gestures to their work along with their subsequent reactions as a response [27]. The takeaway from these and many following research is to comprehend the fact that instead of creating personal or specific user gestures, the development of gesture that can be universally understood, that is, by a large number of people, is the most viable way to go as it includes the concept of gesture creation. While there are numerous studies that include custom gestures [28–31], but most of the mentioned studies do not provide inclusion of gesture creation. Instead, personal gestures are either pre-trained and included in the framework, ready for the end-user to use or provided to user to train of himself/ herself. Multiple areas of application are provided for custom gestures like password, specific actions, or just performing mundane tasks in general.

5.3 Related Fields and Special Mentions

Along with a major literature review about the broad techniques of gesture analysis, this section deals with description and analysis of some latest state-of-art literature in gesture analysis, while also focusing on their specific area of applications.

5.3.1 Collaborative Learning of Gesture Recognition and 3D Hand Pose Estimation with Multi-Order Feature Analysis [32]

While gesture analysis and recognition is itself very abstruse task specially when implemented for feature analysis in hand gestures as various challenges arise as elucidated above, approximating a 3D hand pose in real time opens an altogether new avenue of challenges and problems. This latest state-of-art research study performs 3D hand gesture approximation and recognition with multiple order feature recognition by implementing a novel framework of a network exploring joint aware attributes that appear prominently in all occurring features. It also partly addresses the issue of paucity of data by deliberating the considered task's annotation as a weak supervision. Moreover, to perform both feature and hand pose recognition in 3D poses, it leverages common structures and gestures throughout the hierarchical architecture. This research is very well structured and written with astute recognition to performance and novelty. The framework proposed is well elaborated and explained with appropriate architectural diagrams to assist the same. Based on the results on the grounds of performance, accuracy and space–time complexity, the methodology is phenomenal.

5.3.2 Gesture Analysis and Organizational Research: The Development and Application of a Protocol for Naturalistic Settings [33]

This one-of-a-kind research study performing the very complex yet essential task of analysing, interpreting and deriving conclusion from gestures people use in real time to gather insight about their behaviour, intentions and motives. It includes analysing real-time gestures of people while capturing their gestures and body language via video input to get insights, for example analysing the same from video input of presentations, organization and leadership summits. While the presented work may not be able to provide specific details and exact connotation, it successfully elucidates some general protocols and useful outputs when given different input videos and images. It can make very astute observations about the subject of observation, domain of topic discussed in the input video, general intensity of the environment and formal/informal discussion domain. It classifies gestures in an entrepreneur environment by using verbal or non-verbal indicators. After appropriate training, it succeeds in classifying entrepreneurship and leadership talks into different topics by just analysing gestures of the speaker and his/her body language. Moreover, it can also recognize a particular author by his/her unique style of gestures, body and verbal communication after appropriate training.

5.3.3 Gesture and Language Trajectories in Early Development: An Overview from the Autism Spectrum Disorder Perspective [34]

This study describes a lesser known but substantial domain area of application of gesture analysis and recognition, that is, autism spectrum disorder (ASD) and its close relationship to gesture analysis. Gesture recognition and body language analysis along with nonverbal communication parameters play a cardinal role in early diagnosis of ASD and its regrowth of symptoms. This aims to fulfil two primary purposes in the mentioned domain, namely, verifying the extent of validity of notion that gesture analysis can aid in diagnosis of ASD, and comprehending the prominent differences between the language loss trajectory of various types of autism in early ages of children from various aspects like qualitative aspects, quantitative aspects and more. By the end of this research, the author successfully verifies the validity of the notion and concludes that consistent impairment or delay in response gestures indicates early development of ASD. Additionally, after intensive literature study, it concludes that teaching extra gestures or nonverbal means of communication can not only help these children fulfil the need of social interaction but also build up their confidence to further aid in the treatment of the disorder. After discerning differences in gesture analysis, the research identifies multiple differences between trajectories of frequency of gestures in different stages of ASD, as well differences based on the age of the patient, tracking trajectory can also indicate the course, speed and effectiveness of treatment, if given. Apart from the above, early diagnosis and detection can also be related to the neurological paths that patients follow while their disorder regrows or slips in the dormant state. These insights can be substantial in finding cure to their respective disorders.

5.3.4 Learning Individual Styles of Conversational Gesture [35]

While there are multiple research papers exploring conversational styles, non-verbal gestures, body language and provide variegated models for predicting the subsequent next gestures of their subject of observation after training their model, this research provides ground-breaking results in the domain for multiple speakers. It deliberates speaker-specific data sets of 144-hour video imageries analysing and training in a data-driven technique, to label, clean, pre-process and annotate the data. Post this step, it uses a combination of methods, namely, U-Net architecture for speech to gesture conversion, combination of regression models to predict subsequent motion of the object of observation (speaker in the video) to get the predicted pose sequence, to finally evaluating the obtained results. To further verify the robustness of the proposed algorithm and methodology, Amazon Mechanical Turk is used to conduct a real versus fallacious gesture analysis by providing input to the same. Although the predicted sequence of gestures is respective to the object of observation and are also influenced by the other inter-personal styles that are different for different speakers, this model succeeds in providing high performance for variegated speakers and styles. By the end of study, it provides real-time gesture analysis both qualitatively and quantitatively for data set with path-breaking performance.

5.3.5 Sleep Gesture Detection in Classroom Monitor System [36]

Sleep gesture identification has always been a potential field of application or extended application for gesture analysis but has been comparatively unexplored due to multiple challenges and its demand of being robust considering it has wide real-life application. This research overcomes the limitations of sleep gesture identification like challenge of occlusion, variegated nature of sleep gestures, misidentification of the same with other gestures similar in nature and inability to implement the model in real-life data. The presented state-of-art research collates data from real-life classes of Shanghai, pre-processes the same and implements optimized R-FCN amalgamated with feature pyramid and deformable convolution for gesture recognition. It overcomes specific challenges encountered with real data like recognizing small sleep gestures and also partially identifying loss of attention span of students occurring just before or after sleep gestures. This methodology not only provides ground-breaking performance with noisy video or digital imageries as input.

5.3.6 Towards the Markerless and Automatic Analysis of Kinematic Features: A Toolkit for Gesture and Movement Research [37]

Given the paucity of frameworks that perform robust motion detection, identification of spatial features for real-life data is solved by the presented work. The reason for this is the loss of robustness due to use of hard coding for the same, inability to identify spatial features due to limitation like distance, one at a time approach and limited processing speed. These challenges are very astutely identified and handled by the proposed approach, which captures frequently used or repeated feature of the object of observation, spatial and kinetic attributes. It has wide application and uses due to its robust and dynamic nature.

5.4 Analysis of the Contemporary Literature Pertaining to Gesture Detection

In this section, papers that stand out from the multitude of literature associated in any shape or form with the use of artificial network-based learning techniques for the benefit of humanity are considered. Although the fecundity of the works of the literature mentioned cannot be assimilated in such a small place, each piece of literature is explained in an elaborate manner to provide astute reliable insight into the papers without delving into each one of them individually.

Razieh et al. [38] performed a thorough survey on the current deep learning-based gesture detection frameworks available out there. The chapter starts by listing all the available features used to determine a framework followed by a brief introduction of the parameters. In the next section, all the material collected is sorted based on the type of input for analysis namely, two-dimensional or three-dimensional and the number of colour channels used. Other parameters such as visual modality and the difference between static and dynamic imagery has also been briefly introduced, followed by a chart depiction of all the papers analyses, the corresponding datasets used and the accuracy of the results in percentage or similarity and confidence metrics. This comparison cumulates papers from 2001 to 2019 and includes data sets such as NYU, ICVL, MSRA, LSP, FLIC, ITOP, STB and isoGD, among the notable ones. Multiple discussion sections deal with different aspects of working with models of gesture analysis. The first section discusses the importance of facial features along with suitable examples and elucidates its importance in semantic analysis. This second discussion section deals with the different models usable and their structural importance such as advantages of using ANN over CNN, while the third section discusses the work related to the hybrid approach using the combination of classical classifiers with the deep learning-based descriptors.

Mohammed, in his study [39], reviewed models with a strong focus on the Arabic sign language and reviewed all the pieces of literature that delve into it. Various learning frameworks such as CNN · RNN · MLP · LDA · HMM · ANN · SVM · KNN and SLR systems were explored with elaborations and corresponding accuracies of each of them. One important factor to realize is that unlike its American counterpart, the Arabic sign language systems require both hands to convey some alphabets, which means more spatial awareness is necessary on the parts of the models. Among the 45 papers reviewed, Hartanto's technique scored the lowest with accuracy of 62.8 percent while others scored a mean of 91.2 percent. From these works, it has been concluded that the deep learning-based frameworks rival the analysing capabilities of humans and the neural networks working built on multilayer perceptron possess the highest performance to requirement ratio.

Another real-time application of deep learning in the realm of hand gesture detection was done by Murat et al. [40] on the Tensorflow framework with the Kearas library. The proposed CNN model first converts the coloured image into the YCbCr space to detect the hand using colours. Next, the convex hull algorithm determines the position and location of the hand so that the image can be resized and loaded for classification. These two steps are carried out simultaneously after which a greyscale conversion precluded the feeding into the machine in 30 ms loops. This system was highly capable, scoring 99.44 percent to 100 percent in the validation set. Along with the graphs and images, the real-time working of the system is elucidated.

Hamid et al. [41] created a wearable wrist-mounted device to capture the hand movement data for processing via a deep neural network. Using tri-axial accelerometers and gyroscopes mounted on Inertia Measuring Devices tied to a subject's hands, the passive data during feeding was collected that was further processed to analyse the actions performed. Using a two-step process of identifying intake frames and peak activity frames, the data of 100 individuals was evaluated. After accounting for discrepancies for gravity, the data was processed by the four-layered CNN. According to the results, the system reached a peak accuracy of 88 percent with a varying range. The major drawbacks of the paper come from the system flagging nondeterminate actions like blowing air falsely. The presence of more pooling layers did not drastically change performance neither did changes in the FLOP parameters. The future work refers to the treatment centre of aged patients as a potential beneficiary of such systems.

Andrea et al. [42] designed a LSTM model for the video graphic detection of autism spectrum disorder. The study, done on 20 individuals with a diagnosis of autism, fed a short video clip showing both healthy and autistic participants to the neural network. The experiment consisted of 48 trials of neutral hand gestures of grasping an object, which was filtered for discrepancies. The evaluation for a thresholding value ranging from 0.5 to 0.95 gave subsequent accuracy values of 83.33 percent and recall of 85 percent. The frameworks have been deemed as progressive comparable to state-of-the-art LSTM systems along with the publication of the video data set for public use.

Ahmedt-Aristizabal et al. [43] pursued computer vision to enumerate a facial expression gesture analysis system for the early detection of epilepsy. A system consisting of a video feed recording blinking, chewing, gazing and motion of the jaw at 25 frames, a R-CNN trained with the WIDER data set and a Deep CNN for feature extraction followed by a LSTM with a many-to-one layering for the evaluation. For result generation, a multi-fold cross-validation method was applied, which measures both reliability and consistency of the system. The results discussed show an average accuracy of 96.58 percent with an AUC of 0.9926. This MTLE-based detection has been considered a success by the authors; however, the absence of large public datasets of seizure facial expressions leaves room for overfitting-based bias in the results, as addressed in the paper itself.

Ye [44], in his paper, amalgamates the 3D conventional neural network with the multi-modal feature of the recurrent neural network to gain a superior network. In their experiments, the C2D model is leveraged to extract metric data from short video clips of ASL communication. Using a greedy approach to stitch together the clips with the highest confidence score, the proposed hybrid framework called 3DRCNN can train and work with any of the multitude of ASL present in use. For the data set, the team created a modified Kinect sensor in association with the ASL Association and created short clips of 99 different hand gestures as well as 100 short one-minute video clips of sentences. After training on the Sports1M dataset, testing showed an accuracy of 69.2 percent in case of person-dependent comparison and 65.2 percent in case of person-independent analysis. Both scores were compared and undefeated by two other systems, LRCN and only C3D methods.

In another paper by Rastagoo [45], the team started by building a data set by only 10 participants who provided 10,000 RGB videos of over 90 hand gestures in the Persian language. For the model, a pipeline system starts from a CNN to extract the heatmap of the hand using key points and boundary boxes projected a skeleton of the hand on a plane that feeds it to the proposed framework named 3DCNN. A total of five such parallel streams generates a composite modal joint heatmap, which is processed via a LSTM network. All possible combinations of the data streams with the LSTM, including modifying the number of streams, has been recorded with the best result of 99.80 percent accuracy being crowned to the system of prefiltering with multi-view projection using five streams. Multiple comparisons with the state-of-the-art systems have been made, in which 2DCNN is taken into consideration. In all the cases, the proposed framework scored competitively.

Gomez-Donozo et al. [46] created a CNN-based hand pose approximation system with a mean error less than 5 mm. This system comprising of two CNN is able to accurately predict the position of all joints of a user's hand in the spatial coordinate system from the viewpoint of any camera. The author has further claimed that the system is comparable to a Leap Motion device. The data set used for the training and testing of this system was a novel dataset named *LSMVHandPoses* comprising of 21 sequences in four colours for each of the 20,500 frames, which have been manually annotated and compared with the predicted 2D computerised system. The paper mentions the drawbacks of jitters being introduced in long video clips and suggests to resolve this issue in the future iterations.

Another 3D hand model estimation pipeline was proposed by Spurr et al. [47]. A cross-modal training system composed of semi-supervised learning-based statistical hand model representation in the spatial domain could be used to recreate the position of the hand from the RGB channels of the video feed. For the evaluation, no new data sets were created, and the publicly available data sets Stereo Hand Pose Tracking Benchmark [48] (STB) and the Rendered Hand Pose Dataset (RHD) [49] were used. The paper is too detailed and the intricacies of the KL divergence framework complicated to understand, but the results section is well adorned with over 30 images of detailing the different inputs at different stages for 12 images.

Simon et al. [50] performed two major achievements in their works. First, the main work focuses on creating a fine-tuned full body key point detection system using multiple cameras in tandem, and the corollary of generating an automated key point marking system for automated reconstruction of a music performer. This framework operates using the edge detection techniques of multi-view bootstrap followed by making of keyframes and training a deep neural network. The only evaluation done was comparison of hand gestures, and hand object detection using a Max3 data set against the salient point model proposed in Ref. [51]. Although the framework shows up to 35 percent improvements, the low range of numbers and lack of comparison against other frameworks leaves many avenues unexplored. The system is a novel approach to an already available technique that might be useful in detecting full body gestures in nonverbal communications.

Haque [52], along with his research team, designated a discriminative model for the prediction of the 3D model of a human from one full frame still image. The primary goal of the study was to overcome the drawbacks of the local patch-based spatial arrangement techniques and ultimately achieve viewpoint invariance. An iterative remedial approach was implemented wherein the error feedback would lead to refinement in subsequent passes.

To make the connections more direct with a linear relation to feedback of the previous iterations, the framework used a LSTM as a filtering technique to further reduce temporal dependency and access the tacit layers directly. Evaluation was performed using the public Stanford EVAL data set, which was compared against random forests [53], random tree walks (RTW) [54] and iterative error feedback (IEF) [55] based on the percentage of correct key points (PCKh) metric. The performance was slightly better than the competition and the highest increase in mean scores was found to be in the case of the lower abdomen, while the head gave the least increase according to the metrics. RF was not a suitable choice for contrasting the method as it scored 40 percent lower than both the proposed framework and the other two frameworks, which puts it out of the league of the other frameworks. Another feature of the framework, 'Glimps' added more data with each iteration along the pipeline and provided a noticeable increase in sharpness of the heatmaps presented.

A simple system for feature extraction with gesture detection via a neural network was presented by Jadooki [56]. The distinguishing factor of this paper is that instead of stating all the systems used and the modalities considered, they also provide insightful information about the importance of the difference as well as the effect on results when the parameters are changed. Simple RGB-based image manipulating tools are used to extract the features of the hand and the palm. The effects of lighting and the background intensity has also been mentioned briefly. A simple neural network equipped with a custom-modified discrete cosine transform (DCT) method is used for the feature processing and subsequent result generation. The effect of the multiple layers and the layers that contribute the most to the accuracy of the results have been provided along with graphs of the accuracy over iterations for different tweaks to the variables to the DCT. The results conclude that increasing certain mentioned layers cause linear reduction in the prediction of the hand gesture, till 10,000 iterations of training, after which the framework stagnates.

Priya et al. [57] has proposed the framework of a project to implement an ASL-based text input in laptop systems. This is a bit different from the other studies already discussed, because instead of focussing on the intricacies of the parameters and the pooling of layers of the network, the paper mainly focuses on the requirements of the system to be able to work in real-world applications. Many techniques that are used for the extraction of hand positions implemented deep inside the neural networks such as the convex hull, erosion, black top hat and skeletonize have been studied and the importance evaluated. Finally, the paper suggests methods to implement the gesture as a virtual mouse input via the camera and gives a brief of the theoretical aspects of the project.

The paper drafted by Smrita [58] outlines the use of a radio frequency-based microcontroller-enabled home automation system that can understand gesture commands. The system uses a PIC microcontroller in tandem with an accelerometer that is attached to the wrist as a wearable device to gather the inputs. The microcontroller can converse with various relays using radio frequency, hence there is no need to be in the same room as the appliance that is to be controlled. Currently the system is rudimentary, with the system being able to turn appliances on or off. With proper software support, this system should be capable of carrying out more complex instruction like setting reminders and alarms. The study aims that the system benefits paralytic patients with limited motor functionality, but it can also be applicable to any case where only gesture is the primary means of communication.

5.5 Conclusion

Throughout this chapter, discussions have been made with the primary focus on using gesture detection frameworks that have been proposed, developed and implemented in real-world scenarios. Gesture detection has many uses, starting from the reasons mentioned in the first sections of the chapter to newer intuitive implementation that have not been discovered till date. This is not surprising considering the fact that in this day where digital technology is widespread into every nook and cranny of the society, the access to camera and the ability to use it is extremely trivial. As such, anyone can leverage the fruits of the research done in the fields of gesture analysis, with a main focus still remaining in the domain of 'American sign language', which is one of the most widely used techniques to communicate using palm and hand gestures.

In this chapter, the different challenges with the acquisition and processing of data sets are discussed with certain examples detailing the reason for it. Thereafter, the different requirements along the process of developing usable and efficient frameworks from the system has also been perused over, followed by the application of gesture analysis for the benefit of the society. As previously stated, it is obvious that the most important and obvious use of this technique is in developing tools for communicating with those who are challenged in some form; however, this technique can also be modified for use in novel ways, such as the detection and treatment of autism. Another similar field that got a brief mention was the use in detection of sleep and fatigue levels in students attending online classes as well as making 3D models of complex actions from simple images. There is no doubt that with time, these techniques would get refined by the work of the community, leading to other fruitful discoveries.

Finally, an extended literature review of the current techniques that have something unique to offer have been compiled in the review section. Instead of just stating the work done in those studies, relevant details such as the data sets used, the deep learning techniques implemented for the development of the pipelines, as well as the outcomes and comparative analysis performed have been condensed and mentioned to aid in quick retrieval of important information without access to the original work. A few minor observations have been made: among the techniques evaluated, a large number of them relied on LSTM to finalize and unify the data in the final layer of the models. The use of deep learning-based techniques has seen more prevalence, with normal CNN being used mostly in extracting features from image and video data sets. Most of the studies also had to rely on creating their own data sets, which they have subsequently made public, which can only be seen as a huge asset to the entire research community.

References

[1] Bellugi U, Fischer S (1972) A comparison of sign language and spoken language. *Cognition* 1: 173–200.

[2] Yang R, Sarkar S, Loeding B (2010) Handling movement epenthesis and hand segmentation ambiguities in continuous sign language recognition using nested dynamic programming. *IEEE Trans Pattern Anal Mach Intell* 32: 462–477.

[3] Wu Y, Huang TS (1999) Vision-based gesture recognition: a review. In: International Gesture Workshop, Springer, pp. 103–115.

[4] Wu Y, Huang TS (1999) Human hand modeling, analysis and animation in the context of HCI. In: Image Processing, ICIP 99. Proceedings. 1999 International Conference, IEEE, pp. 6–10.

[5] Starner T, Weaver J, Pentland A (1998) Real-time American sign language recognition using desk and wearable computer-based video. *IEEE Trans Pattern Anal Mach Intell* 20:1371–1375.

[6] Grobel K, Assan M (1997) Isolated sign language recognition using hidden Markov models. In: Systems, Man, and Cybernetics, 1997. Computational cybernetics and simulation. 1997 IEEE International Conference, IEEE, pp. 162–167.

[7] Vogler C, Metaxas D (1998) ASL recognition based on a coupling between HMMs and 3D motion analysis. In: Computer Vision, 1998. Sixth international conference, IEEE, pp. 363–369.

[8] Vogler C, Metaxas D (1997) Adapting hidden Markov models for ASL recognition by using three-dimensional computer vision methods. In: *Systems, Man, and Cybernetics*, Computational cybernetics and simulation. 1997 IEEE international conference, IEEE, pp. 156–161.

[9] Suarez J, Robin R (2012) Hand gesture recognition with depth images: A review. IEEE RO-MAN: The 21st IEEE International Symposium on Robot and Human Interactive Communication Spet.

[10] Ferrone A, Maita F, Maiolo L, and Arquilla M (2016) Wearable band for hand gesture recognition based on strain sensors. IEEE RAS/EMBS International Conference on Biomedical Robotics and Biomechatronics 26–29 June.

[11] Plouffe G, Ana-Maria C. Static and dynamic hand gesture recognition in depth data using dynamic time warping, IEEE Transactions on Instrumentation and Measurement, Vol. 65, no. 2, February 2016.

[12] Wei Lu, Zheng Tong, Jinghui Chu. Dynamic hand gesture recognition with leap motion controller. IEEE signal Processing Letters, Vol 23, No. 9, September 2016.

[13] Alani AA, Georgina Cosma, Aboozar and McGinnity TM. Hand gesture recognition using an adapted convolutional neural network with data augmentation. 4th IEEE International Conference on Information Management, 2018.

[14] Avola D, Bernardi M, Cinque L, Luca Foresti G, Massaroni C. Exploiting recurrent neural networks and leap motion controller for the recognition of sign language and semaphoric hand gestures. *Journal of Latex Class Files*, Vol. 14, No 8, August 2015.

[15] Zabulis X, Baltzakis H, Argyros A. Vision-based hand gesture recognition for human–computer interaction. *Gesture*, 1–56, 2009.

[16] Chakraborty B, Sarma D, Bhuyan M, Macdorman K. Review of constraints on visionbased gesture recognition for human–computer interaction (2018).

[17] Bauer, B., Karl-Friedrich, K. (2002) Towards an automatic sign language recognition system using subunits. In: Wachsmuth, I., Sowa, T. (eds.) GW 2001. LNCS (LNAI), vol. 2298, pp. 64–75. Springer, Heidelberg.

[18] Zhu, Y., Yang, Z., & Yuan, B. (2013, April). Vision based hand gesture recognition. In *2013 International Conference on Service Sciences (ICSS)* (pp. 260–265). IEEE.

[19] Chakraborty, B. K., Sarma, D., Bhuyan, M. K., & MacDorman, K. F. (2018). Review of constraints on vision-based gesture recognition for human–computer interaction. *IET Computer Vision*, 12(1), 3–15.

[20] Liang, RH, Ouhyoung M. (1998, April). A real-time continuous gesture recognition system for sign language. In Proceedings third IEEE international conference on automatic face and gesture recognition (pp. 558–567). IEEE.

[21] Morency, LP, Quattoni A, Darrell T. (2007, June). Latent-dynamic discriminative models for continuous gesture recognition. In 2007 IEEE conference on computer vision and pattern recognition (pp. 1–8). IEEE.

[22] Morency LP, Christoudias CM, Darrell, T. (2006, November). Recognizing gaze aversion gestures in embodied conversational discourse. In Proceedings of the 8th international conference on Multimodal interfaces (pp. 287–294).

[23] Krauss, RM, Chen Y, Chawla P. (1996). Nonverbal behavior and nonverbal communication: What do conversational hand gestures tell us?. In *Advances in Experimental Social Psychology* (Vol. 28, pp. 389–450). Academic Press, London.

[24] Vega K, Cunha M, Fuks H. (2015, March). Hairware: the conscious use of unconscious auto-contact behaviors. In Proceedings of the 20th International Conference on Intelligent User Interfaces, pp. 78–86.

[25] Mitra S, Acharya T. Gesture recognition: a survey. *IEEE Trans. Syst. Man Cybern. Part C Appl. Rev.* 37(3), 311–324 (2007).

[26] Wobbrock JO, Morris MR, and Wilson AD. (2009). User-defined gestures for surface computing. Proc. CHI '09, 1083–1092.

[27] Morris MR, Wobbrock JO, Wilson AD. (2010). Understanding users' preferences for surface gestures. Proc. GI '10, 261–268.

[28] Avrahami D, Hudson SE, Moran TP, Williams BD. (2001). Guided gesture support in the paper PDA. Proc. UIST '01, 197–198.

[29] Bigdelou A, Schwarz L, Navab. (2012). An adaptive solution for intra-operative gesture-based human-machine interaction. Proc. IUI '12, 75–83.

[30] Döring T, Kern D, Marshall P, Pfeiffer M, Schöning J, Gruhn V, Schmidt, A. (2011). Gestural interaction on the steering wheel – reducing the visual demand. Proc. CHI '11, 483–492.

[31] Fiebrink R, Cook PR, Trueman D. (2011). Human model evaluation in interactive supervised learning. Proc, CHI '11, 147–156.

[32] Yang S, Liu J, Lu S, Er MH, Kot AC. (2020, August). Collaborative learning of gesture recognition and 3D hand pose estimation with multi-order feature analysis. In Proceedings of the European Conference on Computer Vision (ECCV).

[33] Clarke JS, Llewellyn N, Cornelissen J, Viney R. (2019). Gesture analysis and organizational research: the development and application of a protocol for naturalistic settings. *Organizational Research Methods*, 1094428119877450.

[34] Ramos-Cabo S, Vulchanov V, Vulchanova, M. (2019). Gesture and language trajectories in early development: An overview from the autism spectrum disorder perspective. *Frontiers in Psychology*, 10, 1211.

[35] Ginosar S, Bar A, Kohavi G, Chan C, Owens A, Malik J. (2019). Learning individual styles of conversational gesture. In Proceedings of the IEEE Conference on Computer Vision and Pattern Recognition, pp. 3497–3506.

[36] Li W, Jiang F, Shen R. (2019, May). Sleep gesture detection in classroom monitor system. In ICASSP 2019–2019 IEEE International Conference on Acoustics, Speech and Signal Processing (ICASSP), pp. 7640–7644. IEEE.

[37] Trujillo JP, Vaitonyte J, Simanova I, Özyürek A. (2019). Toward the markerless and automatic analysis of kinematic features: A toolkit for gesture and movement research. *Behavior Research Methods* 51(2): 769–777.

[38] Rastgoo R, Kiani K, Escalera, S. (2020). Sign language recognition: A deep survey. Expert Systems with Applications, 113794.

[39] Mustafa M. (2020). A study on Arabic sign language recognition for differently abled using advanced machine learning classifiers. *Journal of Ambient Intelligence and Humanized Computing*, 1–15.

[40] Taskiran M, Killioglu M, Kahraman N. (2018, July). A real-time system for recognition of American sign language by using deep learning. In 2018 41st International Conference on Telecommunications and Signal Processing (TSP) (pp. 1–5). IEEE.

[41] Heydarian H, Rouast PV, Adam MT, Burrows T, Collins CE, Rollo ME (2020). Deep learning for intake gesture detection from wrist-worn inertial sensors: The effects of data preprocessing, sensor modalities, and sensor positions. *IEEE Access*, 8, 164936–164949.

[42] Zunino A, Morerio P, Cavallo A, Ansuini C, Podda J, Battaglia F, ... Murino V. (2018, August). Video gesture analysis for autism spectrum disorder detection. In 2018 24th International Conference on Pattern Recognition (ICPR) (pp. 3421–3426). IEEE.

[43] Ahmedt-Aristizabal D, Fookes C, Nguyen K, Denman S, Sridharan S, Dionisio, S. (2018). Deep facial analysis: A new phase I epilepsy evaluation using computer vision. *Epilepsy & Behavior*, 82, 17–24.

[44] Ye Y, Tian Y, Huenerfauth M, Liu J. (2018). Recognizing American sign language gestures from within continuous videos. In Proceedings of the IEEE Conference on Computer Vision and Pattern Recognition Workshops (pp. 2064–2073).

[45] Rastgoo R, Kiani K, Escalera S. (2020). Hand sign language recognition using multi-view hand skeleton. Expert Systems with Applications, 113336.

[46] Gomez-Donoso F, Orts-Escolano S, Cazorla M. (2019). Accurate and efficient 3D hand pose regression for robot hand teleoperation using a monocular RGB camera. *Expert Systems with Applications*, 136, 327–337.

[47] Spurr A, Song J, Park S, Hilliges O. (2018). Cross-modal deep variational hand pose estimation. In Proceedings of the IEEE Conference on Computer Vision and Pattern Recognition (pp. 89–98).

[48] Zhang J, Jiao J, Chen M, Qu L, Xu X, Yang Q. (2017, September). A hand pose tracking benchmark from stereo matching. In 2017 IEEE International Conference on Image Processing (ICIP) (pp. 982–986). IEEE.

[49] Zimmermann C, Ceylan D, Yang J, Russell B, Argus M, Brox T. (2019). Freihand: A dataset for markerless capture of hand pose and shape from single RGB images. In Proceedings of the IEEE International Conference on Computer Vision (pp. 813–822).

[50] Simon T, Joo H, Matthews I, Sheikh Y. (2017). Hand keypoint detection in single images using multiview bootstrapping. In Proceedings of the IEEE conference on Computer Vision and Pattern Recognition (pp. 1145–1153).

[51] Tzionas D, Ballan L, Srikantha A, Aponte P, Pollefeys M, Gall J. Capturing hands in action using discriminative salient points and physics simulation. IJCV, 2016.

[52] Haque A, Peng B, Luo Z, Alahi A, Yeung S, Fei-Fei, L (2016, October). Towards viewpoint invariant 3d human pose estimation. In European Conference on Computer Vision (pp. 160–177). Springer, Cham.

[53] Shotton J, Fitzgibbon A, Cook M, Sharp T, Finocchio M, Moore R, ... Blake A. (2011, June). Real-time human pose recognition in parts from single depth images. In CVPR 2011 (pp. 1297–1304). IEEE.

[54] Yub Jung H, Lee S, Seok Heo Y, Dong Yun I. (2015). Random tree walk toward instantaneous 3d human pose estimation. In Proceedings of the IEEE Conference on Computer Vision and Pattern Recognition (pp. 2467–2474).

[55] Carreira J, Agrawal P, Fragkiadaki K, Malik J. (2016). Human pose estimation with iterative error feedback. In Proceedings of the IEEE conference on computer vision and pattern recognition (pp. 4733–4742).

[56] Jadooki S, Mohamad D, Saba T, Almazyad AS, Rehman A. (2017). Fused features mining for depth-based hand gesture recognition to classify blind human communication. *Neural Computing and Applications*, 28(11): 3285–3294.

[57] Priya, T., Chanukah, P., & Vanusha, D. (2019). Hand gesture to audio based communication system for blind people.

[58] Rumana, T. A. (2015). HAND GESTURE BASED HOME AUTOMATION FOR VISUALLY CHALLENGED. *HAND*, 2(4).

6

Brain–Computer Interface for Dream Visualization using Deep Learning

Brijesh K. Soni[1] and Akhilesh A. Waoo[2]

[1] *Assistant Professor, Department of Computer Science and Technology,*
AKS University, Satna, MP, India

[2] *Head and Associate Professor, Department of Computer Science and Technology,*
AKS University, Satna, MP, India

CONTENTS

DOI: 10.1201/9781003104858-6

6.1 Dream: Introduction

Philosophers around the world believe that dream is one of the most curious things in human beings [1]. The dream is the succession of images and sounds occurring inside the mind during sleep [2]. Researchers have found that visual and auditory stimuli are active during certain stages of sleep within the brain regions [3]. Some researchers found in their research that the hippocampus, amygdala, visual cortex, auditory cortex, and motor cortex are the major regions that are active during the dreaming of a sleeping object [4]. However, most of the oneirologists committed in various articles that dream originates from the centrally located brain stem area but associated with other important cortical areas also.

There are various futuristic research ideas in my mind related to dreams, and belief says that if it will be possible to implement some of them, it might be beneficial for our community. Suppose a dreamer may spend six to eight hours every night sleeping, he is also spending almost an equal time in the workplace during daytime for his survival [5]. So, the question is, is it possible to make use of that time spent sleeping in any way. The dream might be a solution for making sleeping time useful; it means if it is possible to control dream activities and contents, it will help humans to grow financially by increasing our effective work hours.

6.1.1 Theories of Dream

Since the early sixteenth century, scientists are striving to know the fundamental causes of why dream occurs, but still, no one sure about it [6]. There are various philosophical theories proposed by various scientists and philosophers from time to time, and some popular theories of the dream (Figure 6.1) are briefly described here.

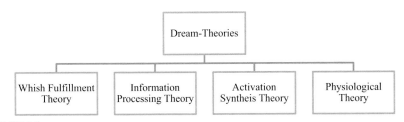

FIGURE 6.1
Various theories of dreaming.

6.1.1.1 Wish-Fulfillment Theory

Wish-fulfillment theory was proposed by an Austrian oneirologist, Sigmund Freud. He believed that dream contents are related to the wishes or desires of the dreamer. Generally, a dreamer sees contents into a dream, which are his incomplete or unfilled desires in his real life [7]. The most common example of such a type of dream is sexual arousal during sleep when an adult dreamer might be involved in sexual or romantic activities in his dream. Some other examples are feeling sadness and happiness, participating in any traditional ceremony, visiting new places or any ancient place and finding voluble things or properties [8].

6.1.1.2 Information-Processing Theory

This theory deals with the process of computation like a computer system to get input from the external environment and process the data and finally generate output. This theory comprises its data transmission path and processes in the human brain just like computer hardware circuits and physical devices are organized in a computer architecture [9].

It is well known that data and information inside the computer system travel through busses and from one device to another device. In the same manner, neurotransmitters travel through the axons and dendrites of neuron cells in the nervous system or biological neural network [10]. Further, the central part of neuron cells, which is also known as soma, is responsible for collecting and modifying information in the form of chemical composition or neurotransmitters. The same concept is also applicable in the case of dream content generation inside the brain [11].

6.1.1.3 Activation Synthesis Theory

This theory is developed based on identifying active brain regions during dreaming. Neuroscientists have seen in experiments that some specific regions are active and produce various types of wave signals, which support various features in dream contents [12]. In-depth, it can be seen that in different regions at different states of mind, various types of signals are generated [13].

6.1.1.4 Physiological Function Theory

This theory of dreaming deals with periodic stimulation of the nervous system. Brain stimulates periodically during the dreaming and strengthens the information stored inside the brain as memory, whatever the dreamer has seen during the waking period [14]. Generally, it happens at every stage of the rapid eye movement (REM) sleep every night. That means the brain works intuitively for strengthening the memory and data pattern inside the brain [15].

6.1.2 Origin of Dream

Scientists undertook imaging of the many regions of the brain that are active while dreaming, starting from the brain stem to the cortex through the thalamus [16]. Most of the observations show that brain regions are active similar to waking stages. Suppose you are seeing any visual imagery during dreaming, it relates to the visual cortex in the occipital lobe of the brain. In the same way, the other cortex is also related to concerned activities [17]. The cortical regions hippocampus and amygdala are responsible for feeling emotions and sensation activities [18].

6.1.2.1 Dream versus Sleep

Sleep is the unconscious state of mind. Sleeping and awakening is a cyclic process in our daily life. Most of the physiologists believe that our whole body, including the brain, needs relaxation, which is possible due to some chemical reactions. That's why every animal takes rest by way of sleeping [19]. Generally, a human being sleeps six to eight hours every night. Further, neuroscientists found that at some stages of dreaming, human brain becomes more active instead of going into a relaxation mode [20]. Various activities are happening during sleeping, but the dream is the most effective activity in our sleeping brain.

Before going into depth about dreaming, a little more about the basics of sleep needs to be discussed. Pieces of literature recognize that sleep can be classified into two broad categories, known as REM sleep and NREM sleep [21]. The REM sleep stage is wholly responsible for dreams occurring in the brain. A dreamer's brain is more active in the REM sleep stage due to dreaming activities (Figure 6.2). Table 6.1 shows time distribution for various sleep stages.

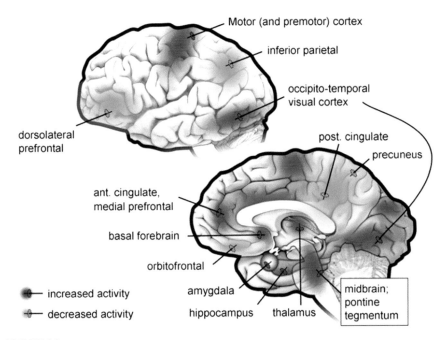

FIGURE 6.2
Neocortical activations during REM sleep.
(*Source*: **www.mentalhealthsciences.com/**)

TABLE 6.1

Time Distribution for Various Sleep Stages

Stages	Episode-1	Episode-2	Episode-3	Episode-4	Episode-5	Total
NREM						
Stage-1	5	5	5	5	5	25
Stage-2	50	40	40	30	20	180
Stage-3	20	20	10	10	10	70
Stage-4	5	5	5	5	5	25
REM	10	20	30	40	50	150
Total	90	90	90	90	90	90×5=450+25*=480

6.1.2.2 Dream versus Memory

Researchers have found that that dreaming is a reasonable activity of memory formation. As we know very well that our brain stores information inside the brain in the form of a chemical pattern, it is further known as memory [22]. Memory is classified into two categories as temporary memory and permanent memory. During dreaming, temporary memory might transform into permanent memory. This process of memory transformation is referred

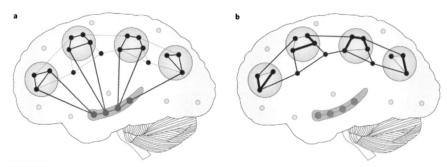

FIGURE 6.3
Long-term memory formation.
(*Source*: **www.nature.com**)

to as memory consolidation or memory synthesis. However, the activation-synthesis theory of dreaming shows that dreaming is also responsible for strengthening memory by analyzing and improving chemical compositions [23]. Initially, temporary memory is formed in the hippocampus, but during consolidation, it might be encoded and stored in the cortical regions of the brain, which might be retrieved in the future. The hippocampus transfers temporary memories to the associated neocortex for long-term storage (Figure 6.3).

Now the importing concept about memory synthesis, which might be a part of memory consolidation while dreaming, is memory encoding. Here the process of memory encoding is discussed with the help of proper construct and pattern. There are four types of memory encoding, discussed in various articles and literature: visual, acoustic, elaborative, and semantic [24]. Stimuli involved in the process of memory encoding may be the dream content a dreamer can see frequently in his real-time dreaming.

6.1.2.3 Human Visual Cortex

The visual cortex is a very crucial part of the human brain located in the occipital lobe. All the visual stimuli are mapped inside the visual cortex. Anatomically, visual cortex is divided into two parts and is located in the left hemisphere and right hemisphere. Here the intentional focus is on the visual cortex because our primary goal is to recognize and visualize visual stimuli only occurring in a dream [25]. Most research engineers and scholars also committed that whatever a dreamer sees in a dream might be mapped within the visual cortex. Even the experimental psychology research outcomes show that the visual cortex is fully active while dreaming (Figure 6.4). However, human curiosity is to know that, if someone got damaged in his visual cortex, then can he see any visual content in his dream? [26].

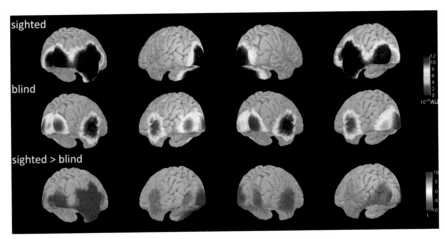

FIGURE 6.4
Visual cortex activation during REM sleep.
(*Source*: **www.nature.com**)

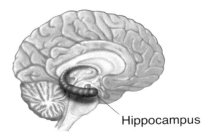

Hippocampus

FIGURE 6.5
Position of hippocampus inside the limbic system.
(*Source*: **https://medimagery.com/**)

6.1.2.4   Hippocampus and Amygdala

Hippocampus is a dominating organ of the human brain where short-term or temporary memory is formed. It regulates motivation and learning, but it plays a crucial role in memory encoding and consolidation also [27]. Physically, it is located in the central area of the brain (Figure 6.5), which means it is a part of the limbic system and thalamus in the cerebral cortex, and strongly connected with all other cortical regions for accessing and forwarding information [28]. Researchers found that the hippocampus plays a vital role in dreaming while a dreamer sees some past experiences and memories in his dream.

Like the hippocampus, amygdala appears in two pieces in each part of the hemisphere (Figure 6.6). It has a basic function in emotional memory [29].

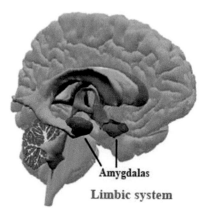

FIGURE 6.6
Position of amygdala inside the limbic system.
(*Source*: www.rebalanceclinic.co.nz)

Researchers believe that like the hippocampus and other cortical regions, amygdala is also active during dreaming and participates in emotional activities in dreaming. A dreamer can feel fear, anger, anxiety, and violence due to the presence of amygdala [30].

6.2   Brain–Computer Interface

At the early stage of this research work, core concepts and anatomical classifications on brain–computer interface (BCI) were published. The present chapter focused on an important application in the domain of dream physiology (oneirology) [31]. The previous section described important brain organs and activities associated with dreaming, and this section discusses some important modalities and techniques frequently used in the domain of brain–computer interface [32].

In clinical bionics, the brain–computer interface is a very popular term among biomedical engineers and researchers. In general, the brain–computer interface is a communication pathway between the human brain and the computer system [33]. Fundamentally, this domain has two broad subdomains where a researcher must focus on modalities and procedures.

- Modalities mean those hardware components which are used to extract signals from the brain invasively or noninvasively.
- The procedure means the data analytics process using software components or modules [34].

Further, these hardware modalities and procedural implementation of software modules are integrally known as brain–computer interface technology. This technology is a very important part of this chapter where it requires to extract dream contents from the human brain and visualize them. First, modalities that can be used may be based on electrical activities or metabolic activity measurements from the brain, and it may be in an individual or hybrid form also [35]. Second, detailed procedure feature extraction and command generation for handling or performing any specific activity is implemented in the form of software modules. Now the point is that these software modules are based on deep learning concepts [36]. However, deep learning theory and respective models with important frameworks are described in the last section of this chapter.

6.2.1 BCI Tools

Researchers are using a tool or modality for extracting signals/images from the brain. Here, some popular modalities are briefly described.

6.2.1.1 Electroencephalography (EEG)

Electrical activity measurement is the most popular study among brain researchers. Electroencephalography (EEG) is one of the major technologies for electrical signal acquisition from the human brain by placing electrodes noninvasively on the scalp during the experiment [37]. EEG has various applications in the domain of brain–computer interface such as medical diagnosis, game playing, skill improvement, and psycho-physiological analysis.

This information can be extracted from various neocortical regions of the brain. As stated in the first section, dreaming is directly associated with memory consolidation happening inside the various neocortical regions of the brain [38]. EEG modality is used for tracking electrical activities occurring during memory consolidation during REM sleep. Electrodes must be placed systematically (10–20 system) in cortical regions. Some major cortical regions are the visual cortex, auditory cortex, motor cortex, and also thalamus region [39]. That means we have to measure electrical activities from the frontal lobe, occipital lobe, and temporal lobe for getting different features of dream contents.

6.2.1.2 Electrocardiography (ECG)

Electrocardiography is also used to measure electrical activities inside the human body, but in general, it is used for recording and analyzing electrical activities of the heart and other associated organs. However, it cannot be used directly for brain diagnosis, but sometimes researchers try to study brain-instructed activities occurring in other associated organs of the body [40].

Psychologists and dream researchers believe that during dreaming, some other organs of the body are also active and may perform some activities other than brain regions. Likely some researchers are actively engaged in dream research using ECG modalities instead of EEG [41]. Nightmares are occur in this type of a dreamer who actively shows body movement during dreaming, and this is directly related to motor activities instructed by the motor cortex [42].

6.2.1.3   Magnetic Resonance Imaging (MRI)

Researchers committed that this is a modern technology for brain imaging compared to EEG/ECG. In this technology, the concept of magnetism is applied instead of electrical activities [43]. Magnetic resonance imaging (MRI) is a radiology-based scanning technique that uses the concept of magnetism, radio waves, and computation to produce images of body structures. This is a tube-like structure surrounded by a large circular magnetic component [44]. The dreamer is placed on a movable bed that is inserted into the magnetic cylinder. The associated magnet creates a strong magnetic field that aligns the protons of hydrogen atoms, which are then exposed to a beam of radio waves [45]. This spins the various regions of the brain, and they produce a faint signal that is detected by the receiver portion of the scanner. A software tool processes the receiver information, which produces an image. The generated image and resolution are quite detailed, and it can detect tiny metabolic changes of structures within the cortical regions [46].

Dream researchers use MRI for detecting active regions within the brain during dreaming. Using the MRI technique, very small active regions can also be detected easily and precisely. Metabolic conditions within the neuron cells show the activities happening during dreaming in REM sleep [47].

6.2.1.4   Magnetoencephalography (MEG)

The basic concept of MEG is also magnetism, but unlike MRI, in this technology any external magnet is not used. It is based on the magnetic property of blood. Magnetoencephalography, or MEG scan, is an imaging technique that identifies brain activity and measures small magnetic fields produced in the brain [48]. The scan is used to generate a magnetic image to pinpoint the source of seizures. In this imaging technique, magnetic fields are detected by extremely sensitive devices [49]. This device is known as superconducting quantum interference device (SQID), frequently used to detect the magnetic field. The MEG scanner is used to detect and amplify magnetic signals produced by the neurons of various active regions of the brain, and unlike MRI, it doesn't emit radiation or magnetic fields [50].

Comparative study of EEG, ECG, MRI and MEG signals acquisition modalities, is shown in Table 6.2.

TABLE 6.2

Comparative Study of Various Signals Acquisition Modalities

Parameter	EEG	ECG	MRI	MEG
Reading	Electrical Activity	Electrical Activity	Structural Imaging	Magnetic Activity
Purpose	Brain diagnosis	Heart diagnosis	Body diagnosis	Brain diagnosis
Channels	Multichannel	Single-channel/ multi-channel	Multichannel	Multichannel
Cost	~2000–3000 USD	~500-1000 USD	~1000000 USD	5000-10000 USD

6.2.2 Procedure

The brain–computer interface is a mediator method and procedure for establishing communication between the dreamers. A BCI communication system has to be built by following various activities and tools such as software and hardware. In this section, a detailed description of developing a BCI system with complete steps is presented.

6.2.2.1 Signal Acquisition

In the formal working procedure of a brain–computer interface system, signal acquisition is the first phase where a hardware modality is used to extract signals from the brain. Generally, hardware modality is associated with the brain directly or indirectly [51]. The way of connection depends on the type of modality; it may be wired or wireless and it may also be invasive, semi-invasive, and noninvasive type of connection. Now consider the example of an EEG machine. So, it may be a wired or wireless system [52]. In early times, researchers used wired EEG systems, but nowadays, most of the dream researchers also used mobile or wireless EEG modality. In the case of EEG-based experiment, an electrode cap having motile electrodes are placed on the scalp. These electrodes read the electrical stimulus from the brain regions, which depends on the number of channels designed in the machine [53].

6.2.2.2 Pattern Recognition

Recorded raw data might have some noise and redundancy, so in this phase, we have to identify patterns or features from the brain signals that have been recorded. Pattern recognition is the task of analyzing the signals to distinguish significant signal features from general raw materials and representing them in a standard form suitable for translation into commands during feature translation [54]. When the input signals to an algorithm are too large and it is suspected to be redundant, then it can be transformed into a reduced set of features. Sometimes, unnecessary portions of the raw data set must be removed, and the process of removing such type of data portion is known as dimensionality reduction. The extracted features are expected to contain valuable and relevant information so that the desired task can be performed by using this reduced pattern instead of the complete source data [55].

6.2.2.3 Pattern Classification

It is the process of categorizing data based on their features to perform the variability reduction of feature values. This stage classifies the extracted feature signals having different features into account [56]. The responsibility of the feature classifier algorithm is to use the feature vector provided by the feature extractor to assign the object to a category of the feature [57]. However, complete classification is often impossible, so a more common task is to determine the probability for each of the possible categories of features. The problem of the classification depends on the variation in the feature values for certain objects in the same category relative to the variation between feature values for certain objects in different categories [58]. The variation of feature values for certain objects in the same category may be due to the complexity of features and may be due to noise in signals.

6.2.2.4 Command Generation

Finally, based on data evaluation, decisions could be taken by computer for performing any specific task by generating appropriate command. This stage performs command generation [59]. This stage translates the signals into meaningful commands for any connected device. The classified feature signals are translated by the feature translation algorithm, which converts the feature signals into the appropriate commands for the specific operations performed by the connected device [60]. In this context source feature, signals are known as the independent variable and targeted device control commands are known as the dependent variable. In the translation process, the independent variable is converted into the dependent variable. Feature translation algorithms may be linear or nonlinear by using statistical analysis and neural networks, respectively [61].

6.3 Deep Learning

Intelligence is a key factor in the human brain; that's why it is known as a super creature of nature. It is well known that intelligence couldn't occur randomly within the muscular tissues, but it is only possible through the continuous process of learning. How a human brain solves problems of recognizing objects and associated patterns even in the case of complex images or high-volume data items [62]. In the same sense, simple algorithms are not capable of solving such types of critical problems so that we need algorithms having the highest capabilities for going into the deepest level and solving such critical problems.

Dreamers around the world believe that mostly dream contents are not easy to interpret and understand clearly. Deep learning provides efficient and successful computational models based on the phenomenon of the human nervous system [63]. These neural network models may be capable of extracting and visualizing dream contents to study comparatively these models with their efficiency and performance [64].

This section provides detailed discussions on various deep learning models and possible frameworks for implementing these models and generating desired results. There are four models in this domain that provide competitive performance with various data sets by adjusting their weights and functions [65]. RNN and CNN are two popular models nowadays among the data scientists for processing and evaluating data sets and generating appropriate outcomes. Gradient-adversarial network (GAN) is another variant of CNN having a hybrid computational capability [66].

6.3.1 Deep Learning Models

Deep learning models are deep neural networks that can be used for processing textual and visual data sets. This section discusses the development of various models for processing and analyzing real-time brain signals during REM sleep. These data might have various features such as auditory data and visual data. Necessary deep neural network models are described in this section.

6.3.1.1 Deep-Belief Network (DBN)

This is the basic structure used for deep learning developed in the early time, which was not more efficient, and now it is replaced by some new and modern models. Deep belief networks are algorithms that use probabilities and unsupervised learning to produce outputs. They are composed of binary latent variables, and they contain both undirected layers and directed layers [67].

Unlike other models, each layer in deep belief networks learns the entire input. In convolutional neural networks, the first layers only filter inputs for basic features, such as edges, and the later layers recombine all the simple patterns found by the previous layers. Deep belief networks, on the other hand, work globally and regulate each layer in order [68].

6.3.1.2 Recurrent Neural Network (RNN)

This is the most popular model for natural language processing. It uses the concept of recurrence using the backpropagation technique for adjusting weights. In this type of network, the output from the previous step is fed as input to the current step [69]. In traditional neural networks, all the inputs and outputs are independent of each other, but in cases like when it is required to predict the next word of a sentence, the previous words are required and hence there is a need to remember the previous words [70]. Thus, RNN came into existence, which solved this issue with the help of a hidden layer. The main and most important feature of RNN is the hidden state, which remembers some information about a sequence [71].

RNN has a memory that remembers all information about what has been calculated. It uses the same parameters for each input as it performs the same task on all the inputs or hidden layers to produce the output. This reduces the complexity of parameters, unlike other neural networks [72].

6.3.1.3 Convolutional Neural Networks (CNN)

This is the topmost neural network model nowadays among researchers and industrialists. This is developed for image and video processing. It can take in an input image, assign learnable weights, and biases to various aspects/objects in the image and be able to differentiate one from the other [73]. The preprocessing required on CNN is much lower as compared to other classification algorithms. While in primitive methods filters are hand-engineered, with enough training, CNN can learn these filters/characteristics [74].

The architecture of a CNN is analogous to that of the connectivity pattern of neurons in the human brain and was inspired by the organization of the visual cortex. Individual neurons respond to stimuli only in a restricted region of the visual field known as the receptive field. A collection of such fields overlaps to cover the entire visual area [75].

6.3.1.4 Gradient-Adversarial Network (GAN)

This is the hybrid model developed by combining two models as a convolutional model and a deconvolutional model. In GANs, there is a generator and a discriminator. The generator generates fake samples of data and tries to fool the discriminator [76]. The discriminator, on the other hand, tries to distinguish between the real and fake samples. The generator and the discriminator are both neural networks and they both run in competition with each other in the training phase [77]. The steps are repeated several times and in this, the generator and discriminator get better and better in their respective jobs after each repetition.

Comparative Study of DBN, RNN, CNN, and GAN Deep Learning Models are shown in Table 6.3.

TABLE 6.3

Comparative Study of Various Deep Learning Models

Parameter	DBN	RNN	CNN	GAN
Type	Staked RBM	Multilayer Perceptron	Multilayer Perceptron	Hybrid Network
Application	Image recognition	String processing	Image processing	Image processing
Vanishing gradient	Solved by chain rule	Suffering from vanishing gradient	Solved by Relu	Solved by Wasserstein loss
Dimensionality reduction	Possible	Possible	Possible	Possible
Algorithm	Unsupervised	Unsupervised	Supervised	Unsupervised

6.3.2 Deep Learning Frameworks

Deep learning is an important element of medical imaging, which includes statistics and predictive modeling. It is extremely beneficial to data scientists who are tasked with collecting, analyzing, and interpreting large amounts of data; deep learning makes this process faster and easier. At its simplest, deep learning can be thought of as a way to automate predictive analytics. While traditional machine learning algorithms are linear, deep learning algorithms are stacked in a hierarchy of increasing complexity and abstraction [78]. Each algorithm in the hierarchy applies a nonlinear transformation to its input and uses what it learns to create a statistical model as output. Iterations continue until the output has reached an acceptable level of accuracy. The number of processing layers through which data must pass is what inspired the label deep [79].

6.3.2.1 Keras

Google Brain developed TensorFlow, which is used as a backend for Keras. Keras is an open-source framework developed by Google engineer Francois Chollet and it is a deep learning framework easy to use and evaluate our models, by just writing a few lines of code [80]. Keras is the best framework to start for beginners, and it was created to be user-friendly and easy to work with Python and it has many pre-trained models—VGG, Inception, among others. Not only ease of learning, but in the backend, it supports Tensorflow and is used in deploying models [81]. Keras was created to be user-friendly, modular, easy to extend, and to work with Python. The API was designed for human beings, not machines, and follows best practices for reducing cognitive load. Neural layers, cost functions, optimizers, initialization schemes, activation functions, and regularization schemes are all standalone modules that can be combined to create new models [82]. New modules are simple to add as new classes and functions. Models are defined in Python code, not separate model configuration files.

The biggest reasons to use Keras stem from its guiding principles, primarily the one about being user-friendly. Beyond the ease of learning and ease of model building, Keras offers the advantages of broad adoption, support for a wide range of production deployment options, integration with at least five backend engines TensorFlow, CNTK, Theano, MXNet, and PlaidML, and strong support for multiple GPUs and distributed training. Further, Keras is backed by Google, Microsoft, Amazon, Apple, Nvidia, Uber, and others [83].

6.3.2.2 Caffe

Convolutional architecture for fast feature embedding (Caffe) is the open-source deep learning framework developed by the University of Berkeley AI Research Group. The framework is available as free open-source software under a BSD license. This framework supports both researchers and industrial applications in Artificial Intelligence [84]. Caffe is a deep learning framework characterized by its speed, scalability, and modularity. Caffe works with CPUs and GPUs and is scalable across multiple processors. Most of the developers use Caffe for its speed, and it can process 60 million images per day with a single NVIDIA K40 GPU [85]. Caffe has many contributors to update and maintain the frameworks, and it is suitable for industrial applications in the fields of machine vision, multimedia, and speech. Caffe works well in computer vision models compared to other domains in deep learning [86].

Caffe can work with many different types of deep learning architectures. The framework is suitable for various architectures such as CNN, LRCN, and LSTM. A large number of pre-configured training models are available to the user, allowing a quick introduction to machine learning and the use of neural networks [87]. As a platform for Caffe come Linux distributions such as Ubuntu but also macOS and Docker container in question. For Windows installations, solutions are also available on GitHub. For the Amazon AWS Cloud, Caffe is available as a preconfigured Amazon Machine Image [88].

6.3.2.3 PyTorch

Nowadays, in the domain of deep learning, PyTorch is one of the popular frameworks among researchers around the world. It is also an open-source framework developed by Facebook AI Research Group; it is a pythonic way of implementing our deep learning models and it provides all the services and functionalities offered by the Python environment. It allows auto differentiation that helps to speed up the backpropagation process [89]. PyTorch comes with various modules like torchvision, torchaudio, and torchtext, which is flexible to work in neural language processing (NLP) and computer vision [90]. PyTorch is more flexible for the researcher than for developers. It provides better performance compared to Keras and Caffe, but it is a low-level API that focused on direct work on array expressions [91].

PyTorch is a Python-based scientific computing package that is a replacement for NumPy to make use of the power of GPUs and a deep learning research platform that provides maximum flexibility and speed [92]. It ensures an easy-to-use API, which helps with easier usability and better understanding when making use of the API [93]. It is fast and feels native, hence ensuring easy coding and fast processing. The support for CUDA ensures that the code can run on the GPU, thereby decreasing the time needed to run the code and increasing the overall performance of the system [94].

6.3.2.4 MATLAB

In the domain of electronics and communication, MATLAB is the most popular software tool since early times. However, it is losing popularity among modern engineers and researchers. A special Deep Learning Toolbox is developed under the umbrella of MATLAB for processing and analyzing a dataset by using a new model, pre-trained models, and apps [95]. However, it provides interfaces for designing and implementing most of the popular neural network models such as CNN, RNN, GAN, and LSTM. Here CNN and LSTM are used to perform classification and regression on image, time series, and text data. Further, GAN and Siamese networks can also be used for automatic differentiation, custom training loops, and shared weights [96]. It uses its programming language and environment, which is different from the Python environment. This tool is originally used for signal processing. However, models can also be imported and exchanged in the Python environment through other popular deep learning tools described in previous sections (TensorFlow, Caffe, and PyTorch) into the MATLAB environment by using the ONNX tool. The toolbox also supports transfer learning with DarkNet-53, ResNet-50, NASNet, SqueezeNet, and many other pretrained models [97].

Deep Network Designer and Experiment Manager are additional applications with MATLAB Deep Learning Toolbox. Deep Network Designer is helpful for designing, analyzing, and training neural networks graphically [98,99]. The Experiment Manager helps in managing multiple deep learning experiments, keeping track of training parameters, analyzing results, comparing code from different experiments, visualizing layer activations, and graphically monitoring training progress [100]. It is easily possible to speed up training using multiple GPU machines or scale up to clusters and clouds, including NVIDIA GPU Cloud and Amazon EC2 GPU instances [101].

Comparative Study of Keras, Caffe, PyTorch, MATLAB Deep Learning Frameworks are shown in Table 6.4.

TABLE 6.4

Comparative Study of Various Deep Learning Frameworks

Parameter	Keras	Caffe	PyTorch	MATLAB (Deep-Learning Toolbox)
Source	Open source	Open source	Open source	Commercial
API	High level	High-level	Low level	Low-Level
Language	Python	Python	Python	MATLAB
Developer	Google Brain	Berkeley University	Facebook	Math-Works
User	Beginners	Engineers	Researchers	Engineers

6.4 Conclusion

This chapter provides just an overview of a dream communication system. However, it's not easy to implement such a type of system, but this is a step toward such significant technology. It may be available among the people in near future. It is considered that information can be processed inside the brain in the same manner as the computer works, while both dreaming and awake. Hence information processing theory is most significant among all other theories of dreaming. Further, it can be seen that during the REM sleep stage, mostly dreams occur in the mind. On the other hand, it can also be seen that most of the memory consolidation happens during the REM sleep stage, which means temporary memory is transferred from the hippocampus to associated neocortical regions for permanent storage, which is also known as long-term memory. Dreaming is the result of memory consolidation, which is a way of learning things by strengthening cortical connections of chemical composition or neurotransmitters. It is possible to read and perform functional activities while dreaming in the REM sleep stage by using various hardware modalities frequently used in the domain of brain–computer interface technology. Finally, we have to map these signals or imageries using deep learning-based models clearly with having better audio and video qualities.

References

[1] Thomas Metzinger, Why are dreams interesting for philosophers? The example of minimal phenomenal selfhood, plus an agenda for future research, *Frontiers in Psychology*, 4:746, 2013.
[2] Walinga and Charles Stangor, *Introduction to Psychology*, BC-Campus Open Education.
[3] Ricardo A. Velluti, Interactions between sleep and sensory physiology, *J. Sleep Res.*, 1997, 6, 61–77.

[4] Julian Mutz, Amir-Homayoun Javadi, Exploring the neural correlates of dream phenomenology and altered states of consciousness during sleep, Neuroscience of Consciousness, Volume 2017, Issue 1, 2017, nix009, https://doi.org/10.1093/nc/nix009

[5] Gwen Dewar, *Sleep Requirements*, Parenting Science Store.

[6] Sander van der Linden, The science behind dreaming, *Scientific American*, 2011.

[7] Sigmund Freud, *The Interpretation of Dreams*, 1900, https://psychclassics.yorku.ca/Freud/Dreams/dreams.pdf.

[8] Wei Zhang and Benyu Guo, Freud's dream interpretation: A different perspective based on the self-organization theory of dreaming, *Front. Psychol.*, Vol 9, 2018, DOI=10.3389/fpsyg.2018.01553.

[9] Eugen Tarnow, How dreams and memory may be related, *Neuropsychoanalysis*, 2014.

[10] Chester A. Pearlman, REM sleep and information processing: Evidence from animal studies, *Neurosci. Biobehav. Rev.*, 1979, 57–68.

[11] Erin J. Wamsley and Robert Stickgold, Dreaming and offline memory processing, *Curr. Biol.*, 2010 Dec 7;20(23):R1010-3. doi: 10.1016/j.cub.2010.10.045. PMID: 21145013; PMCID: PMC3557787.

[12] Michael S. Franklin, The role of dreams in the evolution of the human mind, *Evolutionary Psychology*, January 2005. doi:10.1177/147470490500300106.

[13] J. Allan Hobson, Robert W. McCarley, The brain as a dream state generator: An activation-synthesis hypothesis of the dream process, *The American Journal of Psychiatry*, 1977.

[14] Eiser AS. Physiology and psychology of dreams. Semin Neurol. 2005 Mar;25(1):97-105. doi: 10.1055/s-2005-867078. PMID: 15798942.

[15] Lin Edwards, Dreams may have an important physiological function, *Medical Xpress*, 2009.

[16] Gent TC, Bandarabadi M, Herrera CG, Adamantidis AR. Thalamic dual control of sleep and wakefulness. Nat Neurosci. 2018 Jul;21(7):974–984. doi: 10.1038/s41593-018-0164-7. Epub 2018 Jun 11. PMID: 29892048; PMCID: PMC6438460.

[17] Brown RE, Basheer R, McKenna JT, Strecker RE, McCarley RW. Control of sleep and wakefulness. *Physiol Rev.* 2012 Jul;92(3):1087–187. doi: 10.1152/physrev.00032.2011. PMID: 22811426; PMCID: PMC3621793.

[18] Lukas T. Oesch, Mary Gazea, Thomas C. Gent, Mojtaba Bandarabadi, Carolina Gutierrez Herrera, Antoine R. Adamantidis, REM sleep stabilizes hypothalamic representation of feeding behavior, *Proceedings of the National Academy of Sciences*, 117 (32), 19590–19598, 2020, https://doi.org/10.1073/pnas.192190911.

[19] Corsi-Cabrera M, Velasco F, Del Río-Portilla Y, Armony JL, Trejo-Martínez D, Guevara MA, Velasco AL. Human amygdala activation during rapid eye movements of rapid eye movement sleep: an intracranial study. J Sleep Res. 2016 Oct;25(5):576–582. doi: 10.1111/jsr.12415. Epub 2016 May 5. PMID: 27146713.

[20] Brain Basics: Understanding Sleep. www.ninds.nih.gov/Disorders/Patient-Caregiver-Education/Understanding-Sleep.

[21] *Sleep Physiology*, National Academies Press, 2006.

[22] Memories involve replay of neural firing patterns. www.nih.gov/news-events/nih-research-matters/memories-involve-replay-neural-firing-patterns.

[23] Rasch B, Born J. About sleep's role in memory. *Physiol Rev.* 2013;93(2):681–766. doi:10.1152/physrev.00032.2012

[24] Takeuchi Tomonori, Duszkiewicz Adrian J. and Morris Richard G. M. 2014. The synaptic plasticity and memory hypothesis: encoding, storage and

persistence, *Phil. Trans. R. Soc.* B369:2013028820130288 http://doi.org/10.1098/rstb.2013.0288.

[25] Pisella L, Alahyane N, Blangero A, Thery F, Blanc S, Pelisson D. Right-hemispheric dominance for visual remapping in humans. *Philos Trans R Soc Lond B Biol Sci.* 2011;366(1564):572-585. doi:10.1098/rstb.2010.0258.

[26] Helder Bértolo, Visual imagery without visual perception? *Psicológica* 26 (2005) 173–188.

[27] Chai M. Tyng, Hafeez U. Amin, Mohamad N. M. Saad, and Aamir S. Malik, The influences of emotion on learning and memory, *Front Psychol.* 8 (2017) 1454.

[28] Marcelo R. Roxo, Paulo R. Franceschini, Carlos Zubaran, Fabrício D. Kleber, and Josemir W. Sander, The limbic system conception and its historical evolution, *Scientific World Journal* 11 (2011) 2428–2441.

[29] C. Daniel Salzman, Amygdala. www.britannica.com/science/amygdala

[30] Serena Scarpelli, Chiara Bartolacci, Aurora D'Atri, Maurizio Gorgoni, and Luigi De Gennaro, The functional role of dreaming in emotional processes. *Front Psychol.* 10 (2019) 459.

[31] Albert McKeon, Oneirology is helping clarify our understanding of dreams, https://now.northropgrumman.com/oneirology-is-helping-clarify-our-understanding-of-dreams/

[32] Guangye Li and Dingguo Zhang, Brain–computer interface controlled cyborg: establishing a functional information transfer pathway from human brain to cockroach brain, *PLoS One* 11(3) (2016) e0150667.

[33] Joseph N. Mak and Jonathan R. Wolpaw, Clinical applications of brain–computer interfaces: Current state and future prospects. *IEEE Rev Biomed Eng.* 2 (2009) 187–199.

[34] Scott Makeig, Christian Kothe, and Tim Roger Mullen, Evolving signal processing for brain–computer interfaces, Proceedings of the IEEE 100(Special Centennial Issue): 1567–1584.

[35] Claude Frasson and George Kostopoulos, Brain function assessment in learning. First International Conference, BFAL 2017, Patras, Greece, September 24–25, 2017, Proceedings.

[36] Jason Brownlee, What is deep learning? https://machinelearningmastery.com/what-is-deep-learning/

[37] Lopes da Silva, Fernando. (2010). EEG: Origin and measurement. 10.1007/978-3-540-87919-0_2.

[38] Jessica D. Payne and Lynn Nadel, Sleep, dreams, and memory consolidation: The role of the stress hormone cortisol, *Learning Memory*, 11(6) (2004) 671–678.

[39] Ian G. Campbell, EEG recording and analysis for sleep research, *Current Protocols in Neuroscience.* 2009 October. Chapter: Unit 10.2.

[40] What is an electrocardiogram (ECG)? Institute for Quality and Efficiency in Health Care (IQWiG), 2006.

[41] Kendra Cherry, Why do we dream? www.verywellmind.com/why-do-we-dream-top-dream-theories-2795931

[42] Sara Reardon, Dream movements translate to real life, www.sciencemag.org/news/2011/10/dream-movements-translate-real-life

[43] Magnetic resonance imaging (MRI). www.nibib.nih.gov/science-education/science-topics/magnetic-resonance-imaging-mri

[44] Kathryn Mary Broadhouse, The physics of MRI and how we use it to reveal the mysteries of the mind. https://kids.frontiersin.org/article/10.3389/frym.2019.00023

[45] Magnetic resonance imaging overview, www.pauldurso.com/MRI_technol ogy.htm

[46] Magnetic resonance imaging (MRI), www.cdt-babes.ro/medical-articles/ magnetic-resonance-imaging-mri.php

[47] The brain may actively forget during dream sleep, www.nih.gov/news-events/ news-releases/brain-may-actively-forget-during-dream-sleep

[48] Singh SP. Magnetoencephalography: Basic principles. Ann Indian Acad Neurol. 2014;17(Suppl 1):S107–S112. doi:10.4103/0972-2327.128676.

[49] What is a MEG scanner? www.chop.edu/treatments/magnetoencephalogra phy-meg-scan.

[50] Hamza Alizai, Gregory Chang, and Ravinder R. Regatte. Sample records for field MRI systems, www.science.gov/topicpages/f/field+mri+systems.

[51] Luis Fernando Nicolas-Alonso, Jaime Gomez-Gil, Brain–computer interfaces, a review. *Sensors* (Basel) 12(2) (2012) 1211–1279; Jerry J. Shih, Dean J. Krusienski, Jonathan R. Wolpaw, *Mayo Clin Proceedings* 87(3) (2012) 268–279.

[52] Dahlia Sharon, Matti S Hämäläinen, Roger B.H. Tootell, Eric Halgren, John W Belliveau. The advantage of combining MEG and EEG: comparison to fMRI in the focally-stimulated visual cortex, *Neuroimage* 36(4) (2007) 1225–1235.

[53] Gregory A. Light, Lisa E. Williams, Falk Minow, Joyce Sprock, Anthony Rissling, Richard Sharp, Neal R. Swerdlow, and David L. Braff, Electroencephalography (EEG) and event-related potentials (ERPs) with human participants. *Current Protocols in Neuroscience.* 2010 Jul; Chapter: Unit–6.2524.

[54] Hafeez Ullah Amin, Wajid Mumtaz, Ahmad Rauf Subhani, Mohamad Naufal, and Mohamad Saad, Classification of EEG signals based on pattern recognition approach, November 2017. *Frontiers in Computational Neuroscience,* https://doi. org/10.3389/fncom.2017.00103

[55] Hong Liang, Xiao Sun, Yunlei Sun, Yuan Gao, Text feature extraction based on deep learning: a review, *EURASIP Journal on Wireless Communications and Networking,* Volume 2017.

[56] Jason Brownlee, https://machinelearningmastery.com/an-introduction-to-feat ure-selection/

[57] Anil K. Jain, Robert Duin, and Jianchang Mao, Statistical pattern recogni- tion: A review, *IEEE Transactions on Pattern Analysis and Machine Intelligence* 22(1) (2000) 4–37.

[58] Shahadat Uddin, Arif Khan, Md Ekramul Hossain, and Mohammad Ali Moni, Comparing different supervised machine learning algorithms for disease pre- diction, *BMC Medical Information Decision Making* 19 (2019) 281.

[59] Human–computer interaction, https://psu.pb.unizin.org/ist110/chapter/5-2- human-computer-interaction/

[60] Swati Aggarwal and Nupur Chugh, Signal processing techniques for motor imagery brain–computer interface: A review, *Array,* Volumes 1–2, January– April 2019.

[61] V. Sharma, Machine learning: Introduction to regression analysis, https:// vinodsblog.com/2019/02/01/machine-learning-introduction-to-regression- analysis/

[62] Tim Urban, The AI revolution: Our immortality or extinction, https://waitbut why.com/2015/01/artificial-intelligence-revolution-2.html

[63] Cesar Timo-Iaria (in memoriam); Angela Cristina do Valle, Physiology of dreaming, https://sleepscience.org.br/details/145/en-US/physiology-of-dreaming

[64] Connor Shorten, Taghi M. Khoshgoftaar, A survey on image data augmentation for deep learning, https://link.springer.com/article/10.1186/s40 537-019-0197-0

[65] Deep learning algorithms you should know about, www.simplilearn.com/deep-learning-algorithms-article

[66] Vibhor Nigam, Understanding neural networks. From neuron to RNN, CNN, and deep learning, https://towardsdatascience.com/understanding-neural-networks-from-neuron-to-rnn-cnn-and-deep-learning-cd88e90e0a90

[67] Geoffrey E. Hinton, Simon Osindero, Yee-Whye Teh; A Fast Learning Algorithm for Deep Belief Nets. *Neural Comput* 2006; 18 (7): 1527–1554. doi: https://doi.org/10.1162/neco.2006.18.7.1527.

[68] Deep belief networks: How they work and what are their applications, https://missinglink.ai/guides/neural-network-concepts/deep-belief-networks-work-applications/

[69] Jason Brownlee, A gentle introduction to RNN unrolling, https://machinelear ningmastery.com/rnn-unrolling/

[70] Introduction to recurrent neural network, www.geeksforgeeks.org/introduct ion-to-recurrent-neural-network/

[71] Mahendran Venkatachalam, Recurrent neural networks, https://towardsdata science.com/recurrent-neural-networks-d4642c9bc7ce

[72] V. Sharma, Deep learning: Introduction to recurrent neural networks, https://vinodsblog.com/2019/01/07/deep-learning-introduction-to-recurrent-neural-networks/

[73] Sumit Saha, A comprehensive guide to convolutional neural networks, https://towardsdatascience.com/a-comprehensive-guide-to-convolutional-neural-networks

[74] Shadab Hussain, Building a convolutional neural network, https://towards datascience.com/building-a-convolutional-neural-network-male-vs-female

[75] Convolutional neural network (ConvNet/CNN), www.selfdrivingcars360.com/glossary/convolutional-neural-network/

[76] Jason Brownlee, A gentle introduction to generative adversarial networks (GANs), https://machinelearningmastery.com/what-are-generative-adversar ial-networks-gans/

[77] The discriminator, https://developers.google.com/machine-learning/gan/discriminator

[78] Lee JG, Jun S, Cho YW, Lee H, Kim GB, Seo JB, Kim N. Deep Learning in Medical Imaging: General Overview. Korean J Radiol. 2017 Jul-Aug;18(4):570–584. doi: 10.3348/kjr.2017.18.4.570. Epub 2017 May 19. PMID: 28670152; PMCID: PMC5447633.

[79] Kate Brush and Ed Burns, Deep learning, https://searchenterpriseai.techtarget.com/definition/deep-learning-deep-neural-network

[80] Ketkar, Nikhil. (2017). Introduction to Keras. 10.1007/978-1-4842-2766-4_7.

[81] Purva Huilgol, Pre-trained models for image classification with Python code, www.analyticsvidhya.com/blog/2020/08/top-4-pre-trained-models-for-image-classification-with-python-code/

[82] Gilbert Tanner, Introduction to deep learning with keras, https://towardsdata science.com/introduction-to-deep-learning-with-keras

[83] Martin Heller, What is keras? The deep neural network API explained, www. infoworld.com/article/3336192/what-is-keras-the-deep-neural-network-api-explained.html

[84] Why use keras? https://mran.microsoft.com/snapshot/2018-01-07/web/packages/keras/vignettes/why_use_keras.html

[85] Gautam Ramuvel, Best open source frameworks for machine learning, https://medium.com/coinmonks/5-best-open-source-frameworks-for-machine-learning-739d06170601

[86] What is caffe—The deep learning framework, https://codingcompiler.com/what-is-caffe/

[87] Evan Shelhamer, Deep learning for computer vision with caffe and cuDNN, https://developer.nvidia.com/blog/deep-learning-computer-vision-caffe-cudnn/

[88] Martin Heller, Caffe: deep learning conquers image classification, www.infowo rld.com/article/3154273/review-caffe-deep-learning-conquers-image-classification.html

[89] Caffe Python 3.6 NVidia GPU Production on Ubuntu, https://aws.ama zon.com/marketplace/pp/Jetware-Caffe-Python-36-NVidia-GPU-Production-on-U/

[90] Keras vs PyTorch vs Caffe: Comparing the implementation of CNN, https://analyticsindiamag.com/keras-vs-pytorch-vs-caffe-comparing-the-impleme ntation-of-cnn/

[91] Deep learning frameworks compared: Keras vs PyTorch vs Caffee, https://cont ent.techgig.com/deep-learning-frameworks-compared-keras-vs-pytorch-vs-caffee/articleshow/77480133.cms

[92] Tensorflow vs Keras vs Pytorch: Which framework is the best? https://med ium.com/@AtlasSystems/tensorflow-vs-keras-vs-pytorch-which-framework-is-the-best-f92f95e11502

[93] Implementing deep neural networks using PyTorch, https://medium.com/edureka/pytorch-tutorial-9971d66f6893

[94] Deploying Pytorch in Python via a Rest API with Flask, https://pytorch.org/tutorials/intermediate/flask_rest_api_tutorial.html

[95] Cuda Semantics, https://pytorch.org/docs/stable/notes/cuda.html

[96] Deep Learning Toolbox, www.mathworks.com/products/deep-learning.html

[97] Introducing deep learning with MATLAB, https://in.mathworks.com/campai gns/offers/deep-learning-with-matlab.html

[98] Pretrained deep neural networks, https://in.mathworks.com/help/deeplearn ing/ug/pretrained-convolutional-neural-networks.html

[99] Build networks with deep network designer, www.mathworks.com/help/deeplearning/ug/build-networks-with-deep-network-designer.html

[100] Experiment manager, www.mathworks.com/help/deeplearning/ref/experi mentmanager/

[101] Cloud and data centre, www.nvidia.com/en-in/data-center/gpu-cloud-computing/

7

Machine Learning and Data Analysis Based Breast Cancer Classification

Souvik Das,* Rama Chaitanya Karanam, Obilisetty Bala Krishna, and Jhareswar Maiti

IIT Kharagpur, Kharagpur, West Bengal

E-mail: *rndas9@gmail.com

CONTENTS

7.1 Introduction

The current study preprocesses the cancer patient data, and machine learning models will be trained with this data to classify breast cancer. The study mainly consists of five parts: (i) data collection and understanding, (ii) data preprocessing, (iii) training the ML models, (iv) evaluation, and (v) deployment. We have extracted the required data from the images of cancer cells of 570 patients. We took 30 attributes from the images to classify cancer. The features are computed from a digitized image of a fine needle aspirate (FNA) of a breast mass. They describe the characteristics of the cell nuclei present in the image. For applying machine learning techniques on the data set, we need to preprocess and dispute the data to get a better data set. For analyzing the data set, we need specific Python libraries and packages like Pandas for data manipulation and analysis, NumPy for mathematical operations on the data set, Matplotlib for data visualization, Seaborn for making heatmaps, and Sklearn package for normalization of the data set. In data preprocessing, we remove unnecessary columns from the data frame, change categorical values to numerical values, and standardize the data.

Machine learning is often used to train machines on how to handle the data more efficiently. Here, we have applied supervised learning algorithms to train the model. These include KNN, decision tree, logistic regression, and support vector machine (SVM). The main advantage of using machine learning is that, once an algorithm learns what to do with data, it can do its work automatically. Then we evaluate these models using the Metrics library of Sklearn package. We select the model with high accuracy and deploy it to classify the cancer of the new patient using his cancer cells data.

7.1.1   Review of the Literature

In the literature, it is found that artificial intelligence and Internet of Things (IoT)-based techniques are used to improve the accuracy of the diagnosis of the breast cancer and its prediction. We have provided some of the state-of-the-art machine learning and deep learning approaches, which are used to solve the problem of breast cancer.

Arpit and Aruna [1] optimized a neural network architecture by applying new crossover and mutants and proposed a genetically optimized neural network (GONN) to classify breast cancer.

Ibrahim and Samsuddin [2] proposed a multilayer perception-based neural network model for classifying breast cancer accurately and precisely.

Liu et al. [3] proposed simulated annealing based genetic algorithm and cost-sensitive SVM to improve the classification model for breast cancer diagnosis.

Zemmal [4] presented a computer-assisted genetic algorithm integrated semi-supervised SVM to propose an intelligent breast cancer classification model.

Helwan [5] developed an automated system for the classification of breast cancer based on feed-forward neural network using the back propagation and radial basis function algorithm.

Based on the above-mentioned literature, it can be inferred that more sophisticated machine learning models are to be used for the classification of breast cancer.

7.2   Methodology

Here, we describe a few concepts and algorithms, which we have used for constructing the machine learning models for breast cancer classification.

(i) *Data preprocessing*: The raw data is highly vulnerable to missing data, outliers, noise, and inconsistent data because of the vastness of data, multiple resources, and their gathering methods [6]. The poor quality of data profoundly affects the results of machine learning models. Therefore, preprocessing techniques must be applied to the data to get better results of ML models [7]. After preprocessing, the data must be transformed into the required form for training the ML models [8]. Data preprocessing methods are data cleaning, filling missing values, removing outliers from data, and standardization of data [9]. Missing values of a column data can be filled with the mean of the column data. Outliers can be removed using the binning process. Standardization of data gives a better classification because all the features will be constricted in the range of [−1,1] [10]. We should split the data to train and test data sets in the ratio of 7:3.

Train data set will be used to train the ML model and test data set is used to evaluate the trained ML model [11]. Cross-validation techniques belong to conventional approaches used to ensure good generalization and to avoid overtraining. The basic idea is to divide the data set T into two subsets—one subset is used for training while the other subgroup is left out, and the performance of the final model is evaluated on it. The primary purpose of cross-validation is to achieve a stable and confident estimate of the model performance [12]. Cross-validation techniques can also be used when evaluating and mutually comparing more models, various training algorithms, or when seeking optimal model parameters [13].

(ii) *Training of ML models*: The purpose of machine learning is to learn from the data. Many studies have been done on how to make machines learn by themselves. Many mathematicians and programmers apply several approaches to find a solution to this problem. Some of them are KNN, logistic regression, decision tree, SVM, and random forest. Logistic regression is a classification function that uses a class for building and uses a single multinomial logistic regression model with a single estimator [14]. SVM is the most recent supervised machine learning technique that revolves around the notion of a margin on either side of a hyperplane that separates two data classes. Maximizing the margin and thereby creating the most significant possible distance between the separating hyperplane and the instances on either side of it has been proven to reduce an upper bound on the expected generalization error [14]. Decision trees are those types of trees that groups attribute by sorting them based on their values. The decision tree is used mainly for classification purposes. Each tree consists of nodes and branches. Each node represents attributes in a group that is to be classified, and each branch represents a value that the node can take. In *k*-NN classification, the output is a class membership. An object is classified by a plurality vote of its neighbors, with the object being assigned to the class most common among its *k* nearest neighbors [14].

(iii) *Machine learning algorithms*: Machine learning algorithms, which are used in this study, are explained below.

A. K-Nearest neighbors (KNN): It calculates the distance between the test data point and all the training data points using Euclidean distance formula to check the K nearest neighbors' classes and mode of the k nearest classes will be returned.

Euclidean distance: $\sqrt{(X_1 - Y_1)^2 + (X_2 - Y_2)^2 + ... + (X_n - Y_n)^2}$

B. *Linear regression*: It models a relationship between independent variables and the dependent variable (predictor variable) using a gradient descent approach. After modeling, it can be used to predict the value for a given data. Let's suppose $X_1, X_2, ... X_n$ are the independent variables and $\hat{Y}$ is the predicted value of a dependent variable. Linear regression calculates the weights for each independent variable and returns an equation.

$$\hat{Y} = \alpha + \beta_1 + X_1 + \beta_2 + X_2 + ... + \beta_n + X_n$$

The weights will be calculated using cost function and gradient descent approach.

Cost function:

$$\cos t(\hat{Y}, Y) = \frac{1}{2}(\hat{Y}, Y)^2$$

$$J(\beta) = \frac{1}{n}\sum_{i=1}^{n} \cos t(\hat{Y}_i, Y_i)$$

This $J(\beta)$ function will be plotted in a $(n+1)$ dimensional plane to find the local minimum, that is, to find the values of the weights at which the cost function is minimum. These values will be used for predictions.

C. *Logistic regression*: It uses the sigmoid function to reduce the range of values predicted in linear regression to [0,1]. The output of the sigmoid function is the probability of a data classifying the class mapped to value 1.

Sigmoid function: $\sigma(\hat{Y}) = \dfrac{1}{1 + e^{-Y}}$

Probability that the output is 1: $p(Y = 1 \mid X) = \sigma(\hat{Y})$

Probability that the output is 0: $p(Y = 0 \mid X) = 1 - \sigma(\hat{Y})$

If $\sigma(\hat{Y})$ is greater than or equals to 0.5, logistic regression returns 1 as output, else returns 0 as output.

D. *Decision tree*: A decision tree is a flowchart-like structure in which each internal node represents a "test" on an attribute (e.g., whether a coin flip comes up heads or tails), each branch represents the outcome of the test, and each leaf node represents a class label (decision taken after computing all attributes). The selection of an attribute is based on the entropy of the data and difference in the entropy after selecting an attribute.

Entropy: The randomness of the data set. Here p is the possibility of selecting a class while being randomly picked from a data set:

$$\text{Entropy} = -p(\log(p)) - (1-p)(\log(1-p))$$

Information gain: The information gain is based on the decrease in entropy after a data set is split on an attribute. Constructing a decision tree is all about finding attribute that returns the highest information gain.

$$\text{Information gain} = E_1 - E_2$$

E_1 is entropy of data before splitting.

E_2 is entropy of data after splitting.

E. *Support Vector Machines*: A SVM is a discriminative classifier formally defined by a separating hyperplane. In other words, given labeled training data (supervised learning), the algorithm outputs an optimal hyperplane, which categorizes new examples. In two-dimensional space, this hyperplane is a line dividing a plane in two parts where in each class lay on either side.

SVM uses different kernel functions to map the data into high dimensional plane, where the data can be separated linearly.
Kernel functions used in SVM are the following:

1. Polynomial function
2. Sigmoid function
3. Radial basis function

$$y(x) = \sum_{i=1}^{N} w_i \varphi(\|x - x_i\|)$$

F. *Random forest classification*: It creates *n* different data sets from the training data set and creates a decision tree for each data set. N can be changed manually. Each decision tree will classify the class for the given test data and the mode of the results of *n* decision trees will be given as output.

7.2.1 Evaluation of Machine Learning Models

The metrics we can use to evaluate a machine learning model are accuracy, confusion matrix, logarithmic loss, mean absolute error, mean squared error, precision, and recall. Accuracy is defined as the percentage of correct predictions for the test data. It can be calculated easily by dividing the number of correct classifications by the number of total numbers of testing data. Precision is defined as the fraction of relevant examples (true positives) among all the examples that were predicted to belong to a certain class. The recall is defined as the fraction of examples that were predicted to belong to a class with respect to all of the examples that truly belong in the class. The confusion matrix returns us a matrix, which can be used to describe the performance of a model. The confusion matrix gives the best results in the case of two-class classification problems. The logarithmic loss will be more useful in the case of multiclass classification problems. Accuracy works well only if there are an equal number of samples for each class. F1 score gives a balance between precision and recall. F1 score also can be used to evaluate ml models.

7.2.2 Data Description

The data that is used in the current study for classifying breast cancer contains data of 570 patients of Wisconsin hospital with 32 attributes. The attributes information is given below.

ID number, and Diagnosis (M = malignant, B = benign) are the general attributes.

Ten real-valued attributes are considered and given as follows: (a) radius (mean of distances from center to points on the perimeter), (b) texture (standard deviation of gray-scale values), (c) perimeter, (d) area, (e) smoothness (local variation in radius lengths), (f) compactness (perimeter^2 / area − 1.0), (g) concavity (severity of concave portions of the contour), (h) concave points (number of concave portions of the contour), (i) symmetry, and (j) fractal dimension ("coastline approximation" − 1)

The mean, standard error, and "worst" or largest (mean of the three largest values) of these features were computed for each image, resulting in 30 features.

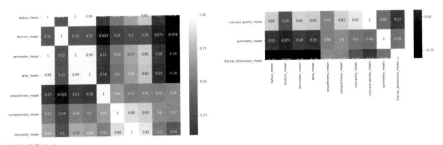

FIGURE 7.1
Correlational plot.

7.3 Heat Map of Correlation

After analyzing the heat map (see Figure 7.1), we found that radius mean, perimeter mean, and area mean are highly correlated. From these, we select one feature for further analysis. We also noted that concave points mean, concavity mean, and compactness mean area highly correlated. In a similar way, we select one feature from these and thus we have removed unnecessary columns and created a new data set.

7.4 Results and Discussions

After preprocessing the raw data, we have applied previously discussed machine learning algorithms on the preprocessed data to classify breast cancer. Models were created and trained in the Jupyter notebook environment and code was written in Python language. The required packages have been installed and used to ease the work. Accuracy score, confusion matrix, precision score, recall score, and f1 score were calculated using Sklearn package. The results are discussed below. Figures 7.2–7.9 provide the variations in accuracy with respect to K-fold for different machine learning algorithms.

(i) KNN

(a) The accuracy of this model is 93 percent with a k value of 7.
(b) Maximum K-fold cross-validation accuracy is 93.75 percent with a k value of 5.

The F1 score of the KNN algorithm is 92.95 percent.

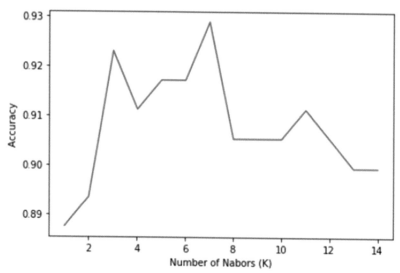

FIGURE 7.2
Accuracy graph for KNN.

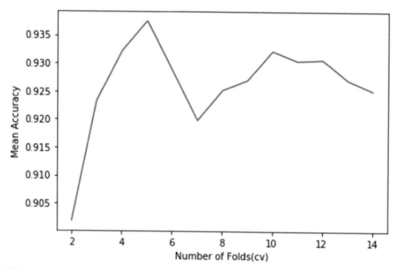

FIGURE 7.3
Cross validation accuracy graph for KNN.

(ii)   Decision Tree

The accuracy is 89.9 percent. Maximum K-fold cross-validation accuracy is 93.6 percent with a k value of 14.

The F1 score of decision tree algorithm is 92.61 percent.

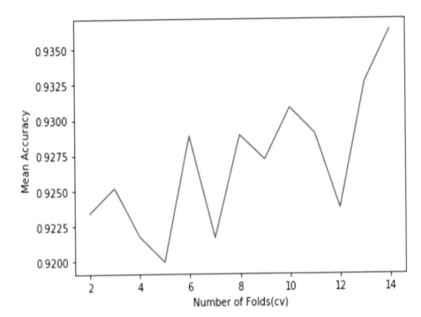

FIGURE 7.4
Accuracy graph for Decision tree.

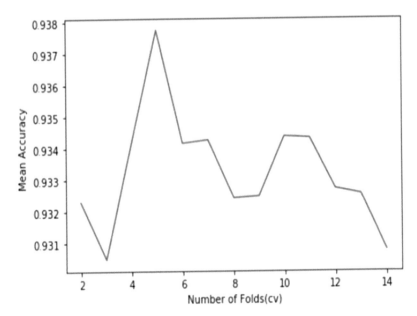

FIGURE 7.5
Accuracy graph for Logistic regression.

(iii) Logistic Regression

The accuracy is 92.9 percent. Maximum K-fold cross-validation accuracy is 93.77 percent with a k value of 5.

The F1 score of logistic regression algorithm is 90.5 percent.

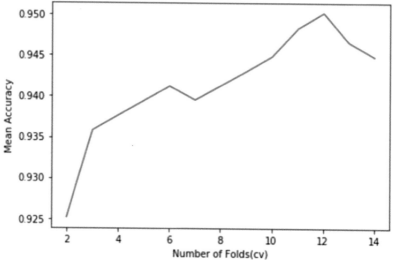

FIGURE 7.6
Accuracy graph for SVM (Rbf).

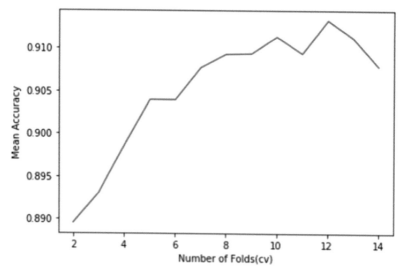

FIGURE 7.7
Accuracy graph for SVM (poly).

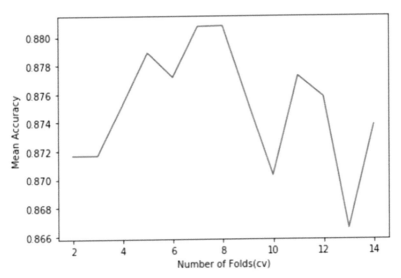

FIGURE 7.8
Accuracy graph for SVM (sigmoid).

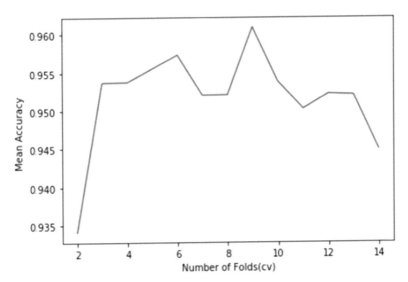

FIGURE 7.9
Accuracy graph for Random Forest.

(iv)   SVM (Rbf)

The accuracy is 92.89 percent. Maximum K-fold cross-validation accuracy is 95 percent with a k value of 12.

The F1 score of SVM (Rbf) algorithm is 90.78 percent.

(v) SVM(Poly)

The accuracy is 89.89 percent. Maximum K-fold cross-validation accuracy is 91.37 percent with a k value of 12.
The F1 score of SVM (Poly) algorithm is 84.61 percent.

(vi) SVM(Sigmoid)

The accuracy is 88.16 percent. Maximum K-fold cross-validation accuracy is 88.07 percent with a k value of 7 or 8.
The F1 score of SVM (Sigmoid) algorithm is 84.5 percent.

(vii) Random Forest

The accuracy is 93.49 percent. Maximum K-fold cross-validation accuracy is 96.08 percent with a k value of 9.
The F1 score of random forest algorithm is 93.5 percent.
From the above results, we can conclude that random forest model is the best model so far with a 96.08 percent accuracy and 0.935 F1 score.

7.5 Conclusion

The current study attempts to solve the problem of classifying breast cancer based on the data obtained from raw images of breast cancer. We have considered various state-of-art machine learning algorithms, such as K-nearest neighbor, decision tree, support vector machine, and random forest. It is found that random forest is the best model for classifying breast cancer. All the experiments and analysis are performed using Python library. This study helps in formulating the process of diagnosing the breast cancer from the preprocessed data obtained from raw images of breast cancer. In future study, deep learning-based tools and techniques can be used to improve the prediction power and make the model more efficient. Future selection process can also be added with the existing study to improve the prediction accuracy.

References

[1] A. Bhardwaj and A. Tiwari, "Breast Cancer Diagnosis Using Genetically Optimized Neural Network Model," *Expert Syst. Appl.*, pp. 1–15, 2015.

[2] A. O. Ibrahim and S. M. Shansuddin, "Intelligent breast cancer diagnosis based on enhanced Pareto optimal and multilayer perceptron neural network," *Int. J. Comput. Aided Eng. Technol.*, vol. 10, no. 5, 2018.

[3] N. Liu, E. Qi, M. Xu, B. Gao, and G. Liu, "A novel intelligent classification model for breast cancer diagnosis," *Inf. Process. Manag.*, vol. 56, no. 3, pp. 609–623, 2019.

[4] N. Zemmal, N. Azizi, N. Dey, and M. Sellami, "Adaptive semi-supervised support vector machine semi supervised learning with features cooperation for breast cancer classification," *J. Med. Imaging Heal. Informatics*, vol. 6, no. 1, pp. 53–62, 2016.

[5] A. Helwan, J. B. Idoko, and R. H. Abiyev, "Machine learning techniques for classification of breast tissue," *Procedia Comput. Sci.*, vol. 120, no. 2017, pp. 402–410, 2018, doi: 10.1016/j.procs.2017.11.256.

[6] M. Han, Jiawei and Jian Pei, and Kamber, *Data Mining: Concepts and Techniques*. Elsevier, Amsterdam, 2011.

[7] J. Han, "Proceedings of the 1996 ACM SIGMOD international conference on management of data," in *Data Mining Techniques*, 1996, p. 545.

[8] Soukup, Tom, and Davidson, Lan, *Visual Data Mining: Techniques and Tools for Data Visualization and Mining*. John Wiley & Sons, 2002.

[9] G. Wang, Jianyong, and Karypis, "On efficiently summarizing categorical databases," *Knowl. Inf. Syst.*, vol. 9, no. 1, pp. 19–37, 2006.

[10] W. S. Alasadi, A. Suad, and Bhaya, "Review of data preprocessing techniques in data mining," *J. Eng. Appl. Sci.*, vol. 12, no. 16, pp. 4102–4107, 2017.

[11] H.R. Bowden, J. Gavin, and Graeme C. Dandy, and Maier, "Input determination for neural network models in water resources applications. Part 1—background and methodology," *J. Hydrol.*, vol. 301, no. 1–4, pp. 75–92, 2005.

[12] W. G. Cochran, *Sampling Techniques*. John Wiley & Sons, 2007.

[13] D. Fernandes, Stenio, Carlos Kamienski, Judith Kelner, and Denio Mariz, and Sadok, "A stratified traffic sampling methodology for seeing the big picture," *Comput. Networks*, vol. 52, no. 14, pp. 2677–2689, 2008.

[14] J. Osisanwo, J.E.T. Akinsola, O. Awodele, J.O. Hinmikaiye, O. Olakanmi, and Akinjobi, "Supervised machine learning algorithms: classification and comparison," *Int. J. Comput. Trends Technol.*, vol. 48, no. 3, pp. 128–138, 2017.

8

Accurate Automatic Functional Recognition of Proteins: Overview and Current Computational Challenges

Javier Pérez-Rodríguez,[1,*] Morteza Yazdani,[2] and Prasenjit Chatterjee[3]

[1]Department of Quantitative Methods, Universidad Loyola Andalucía, Córdoba, Spain

[2]ESIC Business & Marketing School, Madrid, Spain

[3]Department of Mechanical Engineering, MCKV Institute of Engineering, Howrah, India

*jperez@uloyola.es

CONTENTS

8.1 Biological Framework

Bioinformatics is a discipline that binds biology and computer science and deals with tasks such as the acquisition, storage, analysis and diffusion of biological data. The data that this field deals include very often DNA and amino acid sequences. Bioinformatics uses advanced computational approaches to answer a wide variety of problems in molecular biology. Traditionally, the alignment of sequences, gene finding, the establishment of evolutionary relationships or the prediction of three-dimensional forms of sequences were the main problems addressed by this field (Mount, 2004). Recently, the new trends that have appeared include the functional prediction of proteins.

Proteins are complex chemical macromolecules which play a fundamental role in life within organisms (O'Connor et al, 2010). They can be seen, according to their primary structure, as long chains of amino acids connected

DOI: 10.1201/9781003104858-8

to each other. Amino acids are small organic molecules that are linked by a peptide bond. These linear chains fold in space forming a three-dimensional structure. The way of folding is dependent on the amino acids that compose it[1] and determines the properties of the protein.

The importance of proteins lies in the many functions they perform in organisms. For example, at the structural level, they make up the majority of cell material; at the regulatory level, enzymes are proteins; at the immune level, antibodies are glycoproteins. They constitute more than 50 percent of the dry weight of cells (Liu, 2020), and it can be generally stated that they are involved in regulating all the processes that take place in living beings, and in the vast majority of cases, their molecular function is determined by the relationship they establish with other proteins (Braun and Gingras, 2012) and with other molecules in the environment.

Due to how determinant it is to know the function of the different proteins in an organism, many scientific projects are currently focused on trying to elucidate their behaviour, regulation and possible activity from a biochemical, biological, biomedical or bioinformatics angle. Hundreds of thousands of articles are published every year describing the activity of proteins in different situations because it is a crucial factor in analysing cellular mechanisms, identifying the functional changes that lead to possible problems at a systemic level and discovering new tools for the prevention, diagnosis and treatment of diseases. In other words, knowing the function of proteins is synonymous with understanding life at a molecular level, which has a series of relevant implications for the pharmaceutical and biomedical industry.

In late December 2019, the World Health Organization (WHO) was notified of several cases of pneumonia of unknown aetiology, including severe cases. Shortly after, the new coronavirus SARS-CoV-2 (Zhu et al., 2020) was identified as the causative agent. International health authorities recorded a rapid spread of the virus, calling the outbreak of SARS-CoV-2 infections a global pandemic and declaring it a Public Health Emergency of International Importance. Given the seriousness, numerous costly research projects began to be developed around the pathogen. To date, after genome sequencing and analysis, several genes of the coronavirus that causes COVID-19 have been discovered, which encode the synthesis of 39 proteins (Gordon et al., 2020) for the time being. In some cases, it has been possible to identify the functions they play. However, in others, they are still a mystery.

When discussing the function of proteins, it should be borne in mind that this is a concept that can have different meanings depending on the different contexts and/or biological levels. Thus, the biochemical/molecular level, the cellular level and the phenotypic level can be considered. Therefore, the function of a protein can be categorized into three main groups:

- Molecular function
- Biological process
- Cellular component

Nevertheless, the description of the functions of proteins tends to be structured in ontologies, with the purpose of facilitating and standardizing the use and exchange of concepts and knowledge, especially when faced with the problem of massively processing that information. An ontology is a formal definition of types, properties and relationships between entities that exist in a particular domain. In the case of proteins, the activity is described by using hierarchical structures such as directed acyclic graphs or trees. In all these representations, the nodes or terms describe particular activities, and the links represent relationships between the terms. There are different ontology projects that describe the functions of proteins. However, the most popular and widespread is the one developed by the Gene Ontology (GO) Consortium (Gene Ontology Consortium, 2019).

The function of proteins may also vary in time and space. Numerous cases demonstrate this. For instance, alternative splicing (Zheng, 2016) is a well-studied and well-known mechanism in which, after transcription, different exons of the same genome sequence bind in different ways to produce distinct forms of proteins. Similarly, after translation, proteins may be split into functional fragments or the chemical modification of various subgroups (residues) of amino acids can occur, thus changing their properties and, as a consequence, modifying the overall function of the protein. In total, more than four hundred post-translational and post-translational modification events are known, resulting in a very wide range of possible functional forms of a protein, all from the same sequence of amino acids. Another widely known case is that of 'moonlighting' or multifunctional proteins (Jeffery, 2018), proteins that can perform two completely different functions. Furthermore, as mentioned above, on many occasions, proteins interact in groups to carry out separate biological processes. All this has practical implications for the identification and determination of protein functions, which means that this problem remains a significant challenge in the scientific community today.

8.2   Identifying the Protein Functions

Functional annotation of proteins is the set of tasks involved in the process of identifying and assigning labels to the function of known proteins (or groups of them) in organisms. Traditionally, functional annotation was performed through in vitro and in vivo experiments, being recorded in databases of biological sequences through a curation process based on the existing literature. However, these laboratory experiments and manual curation tasks, within a biological context of such complexity as described, were excessively hard in terms of difficulty, inherent costs and time. Advances in technology have led to significant developments in the functional annotation of proteins, but at the same time to considerable growth in the size of biological databases.

Next-generation sequencing (NGS) (Slatko et al., 2018) has boosted a spectacular increase in protein sequence databases as the number of genome and metagenome sequencing projects for many species has grown. It reached a point where the experimental definition of protein function could not handle the vast amount of data of sequences available. In other words, as the production of new collections of sequences increases exponentially, the effort of the teams carrying out in vitro or in vivo experimentation, and of the curators, has not been able in the same way, so that there is a rising number of protein sequences without their supervised or curated annotation being possible. It is, therefore, necessary to develop computational tools that take charge of this work with the greatest possible accuracy. From that moment on, the panorama invited the development of new functional annotation computational techniques.

8.3   Automatic Functional Annotation of Proteins

The simplest and most widely used method to perform the task of automatically determining protein functions is based on homology searches. This is achieved through programs that search for similarities between sequences. For example, Blast2GO (Götz et al., 2008) is one of the most widely used tools of this type, making use of the BLAST algorithm (Camacho et al., 2009) to try to assign to not annotated proteins the function(s) of their homologues, which have been previously inferred and the function is known. However, this analysis depends directly on the databases involved and the annotation of the sequences in them. It is therefore relevant to consider the propagation of the error, that is, if these databases contain errors, the automatic propagation of the errors must be expected, or if the database used is too small, the resulting annotations may be unsatisfactory.

UniProt (UniProt Consortium, 2018) is the consortium that maintains a centralised database that gathers all the functional information on proteins, with precise, consistent and rich annotation. This repository, UniProt Knowledgebase (UniProtKB), is a protein sequence bank consisting of two sections: UniProtKB/Swiss-Prot, which currently contains 562,253 manually annotated and reviewed entries, and UniProtKB/TrEMBL, which contains 180,690,447 entries annotated by automatic, unsupervised, computer techniques and which therefore await full manual annotation.

However, in the situation of an unknown protein that does not have significant similarity to any known protein, computational approaches can be used to predict protein function. These are techniques that use only sequence or structure information to infer properties that are common to proteins of the same function. These methods assume that proteins of the same function adapt similarly to the same conditions. They use information such as the underlying amino acid chain itself (Kulmanov et al, 2018), three-dimensional structure (Yang et al, 2015), binding properties, structural flexibility (Peled et al., 2016), protein–protein interaction networks (Kulmanov et al, 2018) or other molecular and functional factors (Rivarola). Although not directly, these methods may also be partially dependent on protein databases with a known function. Such methodologies rely on supervised machine learning models and can find significant correlations between characteristics and functions.

From a computational perspective, the prediction of protein function can be considered as a multi-label classification problem (Tsoumakas and Katakis, 2007). In the traditional classification machine learning models, each instance is associated with only one class. However, multi-label classification models deal with an environment in which each instance can be associated with more than one class at the same time, which makes multi-label problems more complex and difficult than their single-class equivalents, whether binary or multiple.

Formally, a multi-label problem can be defined as follows (Spolaôr et al., 2013): let D be a data set composed of N examples $E_n=(x_n,Y_n)$, $n=1...N$. Each example E_n is associated with a feature vector $x_n=(x_n^1, x_n^2...,x_n^M)$ described by M features X_m, $m=1...M$, and a subset of labels $Y_n \subseteq L$, where $L=\{y_1, y_2,...,y_q\}$ is the set of q labels. In this scenario, the multi-label classification task consists in generating a classifier H, which, given an unseen instance $E=(x,?)$, is capable of accurately predicting its subset of labels Y, i.e., $H(E) \rightarrow Y$.

The multi-label problems are more complex and difficult than their single-class equivalents, whether binary or multiple. Furthermore, the enormous scale of the problem highly complicates any data-driven approach that aims to address it. Table 8.1 shows the features of the data sets employed for one of the highest performance methods (Kulmanov et al., 2018) currently available. It is clear what the true magnitude of the problem is in terms of its dimension.

At the same time, it should be considered the scenario that arises when dealing with the integration of information from different and heterogeneous sources, in which it is necessary to deal with various biological data (sequence, structure, interactions, etc.). Moreover, a good volume of data is not yet available for most species. That makes the learning process of models with the sequences of one species complicated to apply when it comes to making inferences about others. The limitation of performing a manual validation of

TABLE 8.1

Number of Sequences (patterns) with Experimental Annotations in Data Sets Grouped by Sub-ontologies

	MFO	BPO	CCO	All
Training size	36,110	53,500	50,596	66,841
Testing size	1137	2392	1265	3328
Number of labels	677	3992	551	5220

functional assignments has traditionally led to focusing most of the available resources on species of very high relevance (human, 20,365; common mouse, 17,038; *Arabidopsis thaliana*, 15,952 or *S. cerevisiae*, 6721),[2] which has meant that there is a significant gap in the availability of revised functional data for species considered less essential and/or urgent. Finally, there is not yet a clear procedure on how the performance of annotation methods should be evaluated, since it is necessary to develop good similarity functions between pairs of subgroups in the ontologies.

Functional protein prediction algorithms, both those based on homologies and those based on machine learning, are quite expensive from a computational point of view, both in their development and, especially, in their application. Although the increase in computing power available today, whether CPU-based or increasingly GPU-based, makes it easier to compare new algorithms and to apply them to larger molecules and molecular assemblies. In many cases, the number of sequences in the training set is often limited by the lack of computing resources, which affects the performance of the methods (Rifaioglu et al., 2019).

After an analysis of the results of the predictions of current methods (Figure 8.1) (Radivojac et al., 2013), it can be stated that there is still margin for significant improvement in this field. The critical assessment of functional annotation (CAFA) is an initiative intended for the large-scale evaluation of automatic methods for the functional prediction of proteomes. It is currently in its fourth edition (2019–2020 CAFA4), although the first edition was organised between 2010 and 2011. In the challenge they formulate, they make available to the community a database of protein sequences whose functions are not yet empirically known. This blind data set can be addressed by any computational model to try to predict which functionalities each of the sequences has. Each participant can use the sources of information and procedures they consider appropriate to complete this task. After the closing of the call, an article with the precision of each participating method is made public. The results of CAFA3 (2016–2017) and CAFA 3.14–IP (2017–2018)

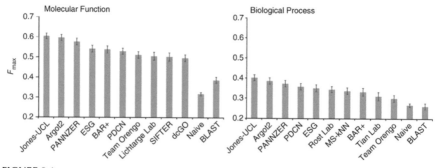

FIGURE 8.1
Overall performance evaluation at two levels.

reveal that homology-based methods are outperformed by those that are not, but that, nevertheless, the measures of accuracy of the best participating models indicate a situation that is far from being able to consider the functional annotation of proteins as a closed problem.

8.4   Challenges

In the light of the aforementioned, after an exhaustive review of the specific literature on the problem of protein function prediction, a number of challenges have been identified that could be of great interest to scientists involved in research projects on this topic. First, to propose new assessment metrics for the problem of automatic functional annotation. Within this point, models have traditionally been evaluated with metrics derived from precision and recall. These metrics are often biased towards the majority classes and do not consider the class hierarchy present in the problem. It would be interesting to propose new evaluation metrics for supervised classifiers implemented to solve automatic protein function prediction problems. Within this area, models have traditionally been evaluated with metrics derived from precision and recall. Such metrics (e.g., *F*-score) tend to have a bias towards majority classes (as does the correct classification rate [CCR]). It also assumes that the distributions associated with the class label distributions and the predicted distributions are equal. In short, the *F*-score is not a suitable metric to measure the goodness of fit of a classifier in a typical bioinformatics scenario with multiple classes and imbalance problems between them. Moreover, this problem presents hierarchical classes so that classification errors should not all have the same weight.

Second, it is considered necessary to develop scalable machine learning models for the problem under study. Functional protein annotation is a classification problem composed of more than 100k instances, more than 5k classes and of a multi-label nature. State-of-the-art machine learning algorithms have serious computational difficulties in addressing problems of this typology due to these models scale up with the sample size of the problem. In order to overcome these serious computational difficulties to deal with problems with so many instances and classes, it would be appealing to test lightweight alternatives (if compared to classical algorithms) and propose models whose computational complexity is adjusted to the needs of the problem. Moreover, these models will have to be designed taking into account that, on the one hand, this is a multi-label problem, and that, on the other hand, they must rely on different and diverse sources of information, which makes an ensemble perspective ideal.

Finally, in line with the above challenge, it should again be noted that the data sets found in the protein function prediction problem contains more than

4,000 attributes. Measures of inter-pattern similarity in machine learning are usually based on the Euclidean metric. This metric has the problem of concentrating distances in high-dimensionality spaces and motivated by this drawback, new inter-pattern similarity non-Euclidean-based metrics need to be proposed for the previously exposed scenario.

Notes

1 The way of folding does not only depend on the sequence of amino acids. For example, prions are proteins that fold in a different way than native protein but do not change their amino acid sequence. There are proteins in the cell that are responsible for the correct folding of the generated new proteins. The physicochemical environment may also affect the final result of the folding.
2 Number of entries in UniProtKB/Swiss-prot.

References

Braun, P. and Gingras, A.C. (2012). History of protein–protein interactions: From egg white to complex networks. *Proteomics*, 12(10), 1478-1498.

Camacho, C., Coulouris, G., Avagyan, V., Ma, N., Papadopoulos, J., Bealer, K. and Madden, T. L. (2009). BLAST+: architecture and applications. *BMC Bioinformatics*, 10(1), 421.

Gene Ontology Consortium (2019). The gene ontology resource: 20 years and still GOing strong. *Nucleic Acids Research*, 47 (D1), D330–D338.

Gordon, D.E., Jang, G.M., Bouhaddou, M. et al. (2020). A SARS-CoV-2 protein interaction map reveals targets for drug repurposing. *Nature* 583, 459–468.

Götz, S., García-Gómez, J.M., Terol, J., Williams, T.D., Nagaraj, S.H., Nueda, M.J., ... & Conesa, A. (2008). High-throughput functional annotation and data mining with the Blast2GO suite. *Nucleic Acids Research*, 36(10), 3420–3435.

Jeffery, C.J. (2018). Protein moonlighting: what is it, and why is it important? *Philosophical Transactions of the Royal Society B: Biological Sciences*, 373(1738), 20160523.

Kulmanov, M., Khan, M. A. and Hoehndorf, R. (2018). DeepGO: predicting protein functions from sequence and interactions using a deep ontology-aware classifier. *Bioinformatics*, 34(4), 660–668.

Liu, S. (2020). *Bioprocess Engineering: Kinetics, Sustainability, and Reactor Design*. Elsevier.

Mount, D.W. (2004). Sequence and genome analysis. *Bioinformatics*. Cold Spring Harbour Laboratory Press: Cold Spring Harbour, 2.

O'Connor C.M., Adams, J.U. and Fairman, J. (2010). *Essentials of Cell Biology*. Cambridge, MA: NPG Education, 1, 54.

Peled, S., Leiderman, O., Charar, R., Efroni, G., Shav-Tal, Y. and Ofran, Y. (2016). Denovo protein function prediction using DNA binding and RNA binding proteins as a test case. *Nature Communications*, 7(1), 1–9.

Radivojac, P., Clark, W.T., Oron, T.R., Schnoes, A. M., Wittkop, T., Sokolov, A. ... and Pandey, G. (2013). A large-scale evaluation of computational protein function prediction. *Nature Methods*, 10(3), 221–227.

Rifaioglu, A.S., Doğan, T., Martin, M.J., Cetin-Atalay, R. and Atalay, V. (2019). DEEPred: Automated protein function prediction with multi-task feed-forward deep neural networks. *Scientific Reports*, 9(1), 1–16.

Slatko, B.E., Gardner, A.F. and Ausubel, F.M. (2018). Overview of next-generation sequencing technologies. *Current Protocols in Molecular Biology*, 122(1), e59.

Spolaôr, N., Cherman, E.A., Monard, M.C. and Lee, H.D. (2013). A comparison of multi-label feature selection methods using the problem transformation approach. *Electronic Notes in Theoretical Computer Science*, 292, 135–151.

Tsoumakas, G. and Katakis, I. (2007). Multi-label classification: An overview. *International Journal of Data Warehousing and Mining (IJDWM)*, 3(3), 1–13.

UniProt Consortium (2018). UniProt: the universal protein knowledgebase. *Nucleic Acids Research*, 46(5), 2699.

Yang, J., Yan, R., Roy, A., Xu, D., Poisson, J. and Zhang, Y. (2015). The I-TASSER Suite: protein structure and function prediction. *Nature Methods*, 12(1), 7.

Zheng, S. (2016). IRAS: High-throughput identification of novel alternative splicing regulators. *Methods in Enzymology* (Vol. 572, pp. 269–289). Academic Press.

Zhu, N., Zhang, D., Wang, W., Li, X., Yang, B., Song, J., ... and Niu, P. (2020). A novel coronavirus from patients with pneumonia in China, 2019. *New England Journal of Medicine*, 382 (8), pp. 727–733.

9

Taxonomy of Shilling Attack Detection Techniques in Recommender System

Abhishek Majumder, Keya Chowdhury, and Joy Lal Sarkar

Department of Computer Science and Engineering, Tripura University, Suryamaninagar, Tripura

CONTENTS

DOI: 10.1201/9781003104858-9

9.1   Introduction

Within the last 20 years, recommender systems have come out as one of the efficient techniques to deal with the knowledge, which is overloaded by suggestive information and it acquires potential interest to the online users. They are helpful to the businesses manufacturing merchandise as it increases the selling rate, cross-sales, and customers' loyalty. As a result, customers tend to come back to the sites that best serve their desires. The recommendation strategies are typically categorized into three main prospects: (i) content-based recommendation, (ii) collaborative filtering, and (iii) hybrid recommendation approaches. Recommendation system, particularly the collaborative filtering (CF)-based system, is introduced with success to filter out irrelevant resources (Si and Li 2020, Sarwar et al. 2001). In the recommender systems, collaborative filtering recommender system (CFRS) is considered as one of the most well-liked and productive techniques. CFRS works on the principle that identical users have identical tastes. However, collaborative filtering results in the source of strength and vulnerability for recommender systems due to its open and interactive nature. Generally, a user-based CA algorithm makes recommendation by searching out similar user patterns, which are illustrated by the preferences of numerous totally non-identical people (Si and Li 2020). If profiles contain biased information, they may be thought as real users and eventually lead to biased recommendations. Therefore, relevant data gets buried under a good deal of irrelevant information. Collective filtering is one in all guidance strategies, which encourage clients to pick a relevant item. It is acclimated for taking care of data overload drawbacks by assembling incredibly right predictions. The key suspicion of collaborative filtering method is that clients having comparable experiences on past things are having a tendency to concur on new things (Bilge et al. 2014a). In any case, these are inclined to profile injection or shilling attacks. As per Lam and Riedl (2004) and O'Mahony (2004), one of the precise negative impacts found in recommender systems (RSs) is that the "deceitful producers" are sometimes used to make a decision for obtaining a more deceitful route. The primary objective of selecting such a route is to influence the RSs in the process of item recommendation considering its rating. Such phenomena can be easily observed from the promotion technique of Sony Pictures for fresh released films in June 2001. The same technique can be used by Amazon.com for selling some non-rated books. Meanwhile, eBay did the same by purchasing good feedbacks (ratings) from other members. Privacy-preserving CF (PPCF) systems have been developed to protect such personal preferences (Bilge et al. 2013). In this chapter, a brief discussion on profile injection technique has been presented. The profile injection technique primarily works on two parameters, namely, rating parameter and rating and time interval parameter.

9.1.1 Collaborative Filtering in Recommender Systems

Collaborative filtering is one-in-all advice technique, which helps users to choose an applicable product. It is accustomed for handling information overload downside by manufacturing extremely correct predictions. The key assumption of collaborative filtering technique is that users having similar experiences on past things tend to agree on new items (Bilge et al. 2014a). Collaborative filtering systems utilize terribly sparse $p \times q$ user-item matrix, which has p users' preferences regarding q merchandise. The systems provide recommendations to their users by evaluating alternative similar users' preference. Collaborative filtering strategies are undefeated at providing correct referrals regarding merchandise. They are additionally able to overcome data overload downside by matching users with right things for them. However, these are prone to profile injection or shilling attacks. There may exist some users or corporations who are malicious aiming to control the recommender systems outcomes for their benefits.

The collaborative filtering strategies are typically categorized into two main prospects: (i) user-based collaborative filtering and (ii) item-based collaborative filtering.

9.1.1.1 *User-Based Collaborative Filtering (UBCF)*

The user-based collaborative filtering (UBCF) basically works on the strategy of recommending items to active users by finding similar users. Figure 9.1

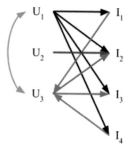

FIGURE 9.1
Framework of user-based collaborative filtering (UBCF).

shows the framework of UBCF. For this technique, it needs two tasks. First, using a similarity function the k-nearest neighbors (kNN) to the user are found. The kNN-based algorithm is the most popular CF algorithm. Data is represented as a $u \times i$ user-item matrix. Where an entry (u, i) indicates either the rating user u gave to item i, if he/she rated it, or null otherwise. Using the Pearson correlation given in Equation (9.1), it computes the similarity between users (Chirita et al. 2005) as

$$W_{ij} = \frac{\sum_{k\in I}\left(R_{ik} - \widehat{R_i}\right)\left(R_{jk} - \widehat{R_j}\right)}{\sqrt{\sum_{k\in I}\left(R_{ik} - \widehat{R_i}\right)^2 \sum_{k\in I}\left(R_{jk} - \widehat{R_j}\right)^2}} \qquad (9.1)$$

where I represents the set of items. R_{ik} and R_{jk} are the ratings users i and j gave to item k. $\widehat{R_i}$ and $\widehat{R_j}$ are the average ratings of user i and j, respectively.

Finally, using the kNN formula, the predictions for user i and item b can be computed as

$$P_{ia} = \overline{R_i} + \frac{\sum_{j=1}^{k} W_{ij}\left(R_{ja} - \overline{R_j}\right)}{\sum_{j=1}^{k} W_{ij}} \qquad (9.2)$$

9.1.1.2 Item-Based Collaborative Filtering (IBCF)

The item-based collaborative filtering (IBCF) algorithm is based on the item–item similarity instead of focusing on the similar user. Figure 9.2 shows the framework of IBCF. This algorithm needs two tasks. First, calculate the similarity between items i and j [8] given in Equation (9.3). There are different types of strategies for finding the similarity among the items. Other similarity calculation techniques are cosine-based similarity, adjusted cosine similarity, correlation-based similarity and 1-jaccard distance.

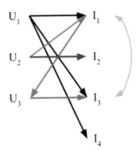

FIGURE 9.2
Framework of item-based collaborative filtering (IBCF).

$$\text{sim}(i,j) = \frac{\vec{i} \cdot \vec{j}}{\|\vec{i}\| \cdot \|\vec{j}\|} \tag{9.3}$$

Then, the prediction for an item is computed by using weighted sum. Here, R_{ub} is a weighted average of user's ratings. Therefore, the prediction score is calculated using the following formula given as

$$P_{ui} = \frac{\sum_{\text{all similar items},a} R_{ua}\text{sim}(i,a)}{\sum_{\text{all similar items},a} |\text{sim}(i,a)|} \tag{9.4}$$

9.2 Profile Injection Attack Method

CFRSs are broadly utilized in e-commerce markets, during which an opponent might try to misguide the recommender framework. Shilling profiles (SPs) are injected by the attacker to the target items just to expand or reduce the recommending frequency (Lam et al. 2004). The shilling profile is also employed as a training sample to work out the recommendations for genuine clients. Therefore, the target items are nominated with additional frequency or else less frequencies (Gunes et al. 2014). Table 9.1 displays a general shilling profile in a shilling attack.

To eliminate detection, SPs could also be visible as legitimate ones. A shilling profile (SP) mainly consists of four elements: (i) the items which are selected (iA) and its rating distribution δ(iA), (ii) filler items (iD) and its rating distribution σ(iD), (iii) the items which are unrated (iα) with null ratings, and (iv) an item which is targeted (iP) and its rating distribution γ(iP) (Bhaumik et al. 2006, 2007, 2011). iA is specified for certain attack method. iD is chosen

TABLE 9.1

General Shilling Profile

Items	iA_1	...	iA_j	iD_1	...	iD_l	$i\alpha_1$	...	$i\alpha_1$	iP
Ratings	$\delta(iA_1)$	...	$\delta(iA_j)$	$\sigma(iD_1)$	...	$\sigma(iD_l)$	Null	Null	Null	$\gamma(iP)$

randomly and attackers carefully style their rates to camouflage as legitimate profiles. iα deals with the non-rated items. iP is the item given the highest or lowest rating. It also indicates the item which the offender aims to attack. The strength of shilling attack (SA) is described by the size of the filler and size of the attack. The ratio of range of the items, which are rated in a profile to the whole variety of items, represents the size of the filler. The ratio of the quality of SP and the total number of all profiles denotes the size of the attack. The size of the filler and size of the attack are usually not large for the price of attack and to ignore detection.

9.3 Shilling Attack Detection

To reduce the effect of shilling attacks (SAs) in CFRSs, there are two common techniques (Si and Li 2020). One technique is SA detection and other technique is before running CF algorithms excludes the attack profiles from the rating information. Several schemas are proposed to find such SAs. Another different approach is to develop attack-resistant CF algorithm. The SA identification algorithms will be classified as follows: supervised classification techniques, unsupervised clustering techniques, semi-supervised techniques, and other techniques.

9.4 Rating

Rating is a measurement of the quality of something, especially when compared with other things of the same type.

9.5 Time Interval

A clock breaks the time into the intervals of hours, minutes, and seconds. An interval is a discrete measurement of time between two things.

9.6   Classification

In this section, a classification of shilling attack detection technique has been presented. Based on the parameter used, the techniques have been classified into two types: (i) using rating parameter and (ii) using rating and time interval parameter. The techniques are further classified into three types based on the output: (i) attack profile detection, (ii) attack item detection and attack profiles, and (iii) attack item detection. The classification is shown in Figure 9.3.

9.6.1   Using the Rating Parameter

In this type of technique, the shilling attack is detected using rating parameter.

9.6.2   Detection of Attack Profiles

This type of technique is used to detect attack profiles. Examples of this type of techniques are Hilbert–Haung transform and support vector machine

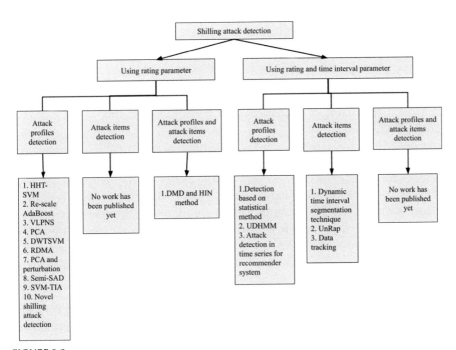

FIGURE 9.3

Classification.

(HHT-SVM), re-scale AdaBoost, variable-length partitions with neighbor selection (VLPNS), principal component analysis (PCA), discrete wavelet transform and support vector machine (DWTSVM), PCA and perturbation, semi-supervised learning using the shilling attack detection (SEMI-SAD), support vector machine and target item analysis (SVM-TIA), rating deviation from mean agreement (RDMA), and novel shilling attack detection.

9.6.2.1 Hilbert–Haung Transform and Support Vector Machine (HHT-SVM)

The Hilbert–Haung transform and support vector machine (HHT-SVM) technique (Fuzhi and Zhou 2014) has been employed to identify the profile injection attacks by using the combination of support vector machine and Hilbert–Haung transform. The HHT is accustomed to transforming the rating series of the individual user profile for characterizing the profile injection attack. The SVM is used to detect profile injection attack. The disadvantage of this method is as soon as new types of attacks occur, the SVM classifier requires being re-trained offline. It is necessary for HHT-SVM to know about new kind of attacks incrementally, because in real CFRs' context, attacks are changing dynamically.

9.6.2.2 Re-scale AdaBoost

This detection method improves the overall performance with respect to following two aspects. Initially, it finds out on well-structured features from client properties. Then using the statistical properties based on the various attack models, it makes hard identification circumstances to act simpler. Later, with reference to the general thought of re-scale Boosting (RBoosting) and AdaBoost, it tends to use a variation of AdaBoost, known as the re-scale AdaBoost (RAdaBoost) (Yang et al. 2016). The RAdaBoost detection technique supports the extracted features. This implies that if the parameters are selected suitably, the RBoosting technique is basically an optimal boosting-type calculation. The RAdaBoost may be employed in conjunction with several alternative sorts of the machine learning (ML) algorithms for upgrading the capability of SA recognition. This technique also increases stress on concerned attacks. On a troublesome classification task, it will clearly improve the predictive capability.

9.6.2.3 Variable-Length Partitions with Neighbor Selection

Variable-length partitions with neighbor selection (VLPNS) algorithm (Wei and Shen 2016) is designed to reduce the matter of unassured prediction accuracy of CFRS techniques against the SAs. It relies upon threshold setting. VLPNS has the particular property of using soft decision procedure that holds suspicious fakers instead of erasing them. It enables the consideration of mistakes of the traditional user during similarity calculation. Neighbor choice with variable-length allotments create segments with entirely unexpected sizes by c-means grouping. It picks neighbors, which are consistent with each client's suspicious possibility. The technique diminishes the likelihood of legitimate clients being detected as fakers. In this way, the prediction accuracy is increased.

9.6.2.4 Principal Component Analysis

Detecting shilling attacks improves the security of a recommender system (Lam S K et al. 2004). Principal component analysis (PCA) detection (Deng et al. 2016) is designed to detect shilling profiles in a recommender system (Mehta B 2007; Mehta B et al. 2007; Mehta B and Nejdl W 2009). PCA can be defined as a direct dimensionality decrease method, which generally intends to project data over another space having lower dimensionality as well as little co-variance among the dimensions. High co-variance among the dimensions is often viewed as excess information and they need to be removed. So, the repetition of information is decreased in the original data at substitution space (Jolliffe 2011). PCA is generally used for finding out shilling profiles. PCA views clients like dimensions of a client-item matrix. In this manner, a profile consisting of very little linear combination of weight has additional probability of being a shilling profile. PCA detection acts well enough for identifying the shilling profiles, which are created using attack models like average attack, bandwagon attack, and random attack (Mehta B 2007).

9.6.2.5 Discrete Wavelet Transform and Support Vector Machine (DWTSVM)

The detection technique is basically a hybrid approach by merging the discrete wavelet transform and the support vector machine (DWTSVM) (Karthikeyan et al. 2016). To categorize fake or legitimate clients, the DWTSVM method constructs ratings of each client by utilizing the popularity and novelty of items rated by the clients. In this method, some nonstationary signals established by employing the popularity and novelty of items the clients are rated, which is given as an input to the DWT. The instantaneous frequency, amplitude and phase of DWT signal are applied to a feature set. Then it is used by the SVM for classifying the users as fake or genuine users.

9.6.2.6 Principal Component Analysis and Perturbation

The PCA and perturbation (Deng et al. 2016) is an unsupervised shilling attack (SA) detection technique. PCA is generally applied at the beginning and after application of Gaussian noise in every user profile. Therefore, the SAs are discovered by merging the results obtained from two PCAs. This technique achieves higher accuracy in experiments as rating of a shilling profile (SP). The SP contains less deviation from average rating. The effect of injecting perturbation to SPs ought to be greater than that of the legitimate profile. The experiment result confirms that this technique outperforms the initial PCA. Injecting perturbation is useful for shilling attack detection.

9.6.2.7 Semi-supervised Learning Using the Shilling Attack Detection (SEMI-SAD)

The semi-supervised learning based shilling attack detection (SEMI-SAD) technique (Wu et al. 2011) is assisted by shilling attack detection algorithm. It takes advantage of each kind of information. At first, it trains a small set of labeled clients as naive Bayes (NB) classifier and then consolidates the unlabeled clients with expectation maximization (EM) for upgrading the said NB classifier. The semi-SAD can more readily distinguish numerous sorts of shilling attacks compared to others, particularly obfuscated and hybrid shilling attacks. This technique optimizes naive Bayes in form of the initial detector and an augmented EM to enhance the detector. This algorithm gains knowledge from each labeled and unlabeled client profiles. By combining naive Bayes classifier and unlabeled data, EM is utilized to increase the weight of the labeled information.

9.6.2.8 Support Vector Machine and Target Item Analysis (SVM-TIA)

This shilling attack detection technique (Zhou et al. 2016) employs support vector machine and targets item analysis in the recommender systems. It works on Border line-SMOTE strategy, which is utilized to reduce the class unbalance drawback in taxonomy. Therefore, the fine-tuning part, which is used to target items within the potential attack profile set, is analyzed. This system analyses the ratings of the items to observe the attack profiles. However, the disadvantage of the technique is that some genuine profiles are misjudged as an attack profile.

9.6.2.9 Rating Deviation from Mean Agreement

The rating deviation from mean agreement (RDMA) technique (Si and Li 2020; R. Burke et al. 2006) observes attackers though testing of average deviation of the profile for each item. It is weighted by the inverse of the amount of rating of the item to detect the attack profiles.

9.6.2.10 Novel Shilling Attack Detection

The novel shilling attack detection (Bilge et al. 2014b) method is used for specific attacks. It uses bisecting k-means clustering approach, which allocates a binary decision tree (BDT) where attack profiles are gathered in a leaf node. BDT is constructed by recursively clustering the training data to locate the fake attack profiles using k-means clustering algorithm. The user-item matrix is divided into two distinct clusters at each level. Intra-cluster correlation coefficient is calculated for each internal node. The process is repeated until there remains at most a predefined number of users in any leaf node. Then BDT is traversed to detect anomalies with intra-cluster correlation coefficient and label the node holding all or most of the attack profiles.

9.6.3 Attack Items Detection

No technique has been introduced yet, which is designed for attack items detection.

9.6.4 Attack Profiles and Attack Items Detection

This type of technique is designed to detect attack profiles and attack items. Examples of this type of techniques are double M detector (DMD) and heterogeneous information network (HIN) method.

9.6.4.1 Double M Detector (DMD) and Heterogeneous Information Network (HIN) Method

The double M detector (DMD) technique detects shilling attack by meta-path and matrix factorization. The DMD technique (Zhang et al. 2019) concatenates the client-item bipartite network and client–client relation network. Then it designs many meta-paths, which control the stochastic procedure to produce node sequence and uses the skip-gram model to get user embedding. To get the latent factors, matrix factorization is utilized to decompose the client-item rating matrix. For representing users' embeddings of latent relations, numerous significant meta-paths have been designed using heterogeneous information network (HIN) with respect to network characteristics, such as coreness, degree, and hindex. At the end, exploitation embedding and factors are utilized together to train the identifier to detect the attack profiles and attack items.

9.6.5 Using the Rating and Time Interval Parameter

In this type of technique, the shilling attack is detected using rating and time interval parameter.

9.6.6 Attack Profiles Detection

This type of technique is used to detect attack profiles. Examples of this type of techniques are detection based on the statistical model, UD-HMM, and attack detection in time series.

9.6.6.1　Detection Based on the Statistical Model

The statistical detection algorithm (Wang et al. 2018) first segments the features of every item with respect to its time-ordered rating sequence. Statistical distributions are designed to get the suspicious rating's relating time interval. This technique is often drawn as hypothesis test detection and dynamic time interval segmentation based strategy (Xia et al. 2015). Based on the central supposition that successful attacks ought to alter the target item's statistical properties, this identifier wants to get the irregularity changes of items rating sequence. The hypotheses test is employed after segmentation of the rating series just to check irregular user groups among all the groups.

9.6.6.2　Unsupervised Technique Using the Hidden Markov Model and the Hierarchical Clustering

The shilling attack identification technique relies on the hidden Markov model (HMM) and hierarchical clustering (Zhang et al. 2018). This technique measures the difference of user rating behaviors by constructing every user's rating item sequence. Then it uses HMM model to get each user's preference sequence for calculating each user's matching degree. It calculates each item's entropy to find the users' suspicious degree for detecting the attack profiles. The limitation of this technique is that the detection performance isn't excellent while detecting the collusive spammer on the sampled data set.

9.6.6.3　Attack Detection in Time Series

The attack detection in a time series technique (Zhang et al. 2006) uses a window of size k for building a time series of an item's rating. It clusters successive rating of items into disjoint windows. Then it determines the sample entropy and sample average of the window. It derives a theoretical optimal size of window for best observation of an attack event if amount of attack profiles is known. This work begins with the subsequent observation. It assumes that the attack profiles are infused into the framework in a comparatively less amount of time. Most of the shilling attack models share a common trend in spite of decent variety of the attack amount they induct in the target items rating distributions (probably for different items). Moreover, in a large-scale recommender framework, the ratings of a given item may have seasonality and/or pattern over time. A lot of complicated models are required to include such trends.

9.6.7 Attack Items Detection

This type of technique is used to detect attack profiles. Examples of this type of techniques are dynamic time interval segmentation technique, unsupervised retrieval of attack Profiles (UnRAP), and data tracking.

9.6.7.1 Dynamic Time Interval Segmentation Technique

The dynamic time interval segmentation technique (Xia et al. 2015) is employed for locating out items attacked by malicious profiles directly. The entire framework of this technique contains two parts, that is, time interval segmentation and abnormal interval detection. The time interval segmentation technique segments the life cycle of every item into many time intervals dynamically at a checkpoint. The abnormal interval detection method detects an anomalous item through hypothesis testing. This detection technique improves the detection performance against target shifting attack.

9.6.7.2 Unsupervised Retrieval of Attack Profiles

This unsupervised retrieval of attack profiles (UnRAP) detection technique (Si and Li 2020) is an unsupervised method. This technique detects shilling attack by analyzing user profiles rating deviation on target time. This technique can solely detect the attack user profiles of individual time.

9.6.7.3 Data Tracking

This data tracking detection technique (Qi et al. 2018) is used in huge information atmosphere. This technique supports new information feature. The technique uses extended Kalman filter, which quickly tracks and accurately predicts the rating status of the item supported two new detection attributes, short-term average change activity (SACA) and short-term variance change activity (SVCA). The detector then detects the abnormal item by comparing predicted ratings and actual ones.

9.6.8 Attack Profiles and Attack Items Detection

No technique has been introduced yet, which is based on detection of attack profiles and attack items as output.

9.6.9 Advantages and Disadvantages

In this section, the advantages and disadvantages of the shilling attack detection techniques have been presented in Table 9.2.

TABLE 9.2

Advantages and Disadvantages

The shilling attack detection technique	Advantages	Disadvantages
1. Novel shilling attack detection (Bilge et al. 2014b).	1. This method gives promising results for almost all filler size and attack size.	1. This method detects only specific three types of attacks namely bandwagon, segment, and average attack. 2. When all the profiles are genuine in the system, this technique misjudges the genuine profiles as malicious profiles.
2. Hilbert–Haung Transform and support vector machine (HHT-SVM) (Fuzhi et al. 2014).	1. This method decomposes each rating series. It also extracts Hilbert spectrum-based features for characterizing the profile injection attacks. 2. Here SVM distinguishes the genuine users' profile and attacker profile.	1. In this method, as soon as new types of attacks occur, the SVM classifier requires to be re-trained offline.
3. Re-scale AdaBoost (Yang et al. 2016).	1. This method extracts well-designed features from user profiles. For improving detection performance, these profiles are established based on the various attack models' statistical properties. 2. It makes hard detection scenarios easier to perform.	1. The detection rate of this method is low for the attacks that have small size of attack and filler size. 2. Re-scale AdaBoost cannot effectively detect Power User Attack-Aggregate Similarity (PUA-AS), Power User Attack-Number of Ratings (PUA-NR) and Power User Attack-In Degree (PUA-ID) attacks. 3. Generic features and type-specific features which are present in this technique as extractive features are not enough to depict their material characteristics.
4. Rating Deviation from Mean Agreement (RDMA) (Si and Li 2020; Burke et al. 2006).	1. This algorithm successfully detects attack profiles which are random, average and bandwagon.	1. It is unable to detect segment attack and love/hate attack.

TABLE 9.2 (Continued)
Advantages and Disadvantages

The shilling attack detection technique	Advantages	Disadvantages
5. Variable-length partitions with neighbor selection (VLPNS) (Wei and Shen 2016)	1. Instead of detecting the suspicious fakers, this technique marks them and thereby reduces false-positive rate. This process is carried out in such a way that the misclassified normal users will be able to still contribute to the similarity calculation. As a result of this, the probability of mistaking normal users as fakers reduces. 2. The prediction accuracy of this method is high.	1. VLPNS has a higher time complexity. 2. This algorithm does not improve the false negative.
6. Principal component analysis (PCA) (Deng et al. 2016).	1. PCA performs very efficiently against standard attacks. 2. This detection method performs very well to identify shilling profiles in bandwagon attack, random attack, and average attack models.	1. The applicability of PCA is limited by certain assumptions made in its derivation. 2. Another limitation for this method is the mean-removal process before constructing the co-variance matrix for PCA.
7. Discrete wavelet transform and support vector machine (DWTSVM) (Karthikeyan et al. 2016)	1. This method forms rating series of individual users. The rating series is formed with reference to the popularity and novelty of items. This classifies the users as fake or genuine users based on time, so that the detection rate is easy. 2. For getting the feature set, discrete wavelet transform is applied on the rating series. It is further used by support vector machine for the purpose of classification.	1. It applies Nyquit's rule. As a result, half of the signal frequencies are ignored while sending through the filters. 2. DWT is computationally intensive. It is less efficient and natural. 3. The dimensions of SVMs may be very high.
8. Principal component analysis and perturbation (Deng et al. 2016).	1. By utilizing PCA detection technique, the attack and legitimate profiles can be separated out. 2. This method is effective, simple, and can be easily implemented.	1. In this method the applicability of PCA is limited by certain assumptions made in its derivation. 2. When perturbation accuracy is decremented, the state values have higher complexity.

(continued)

TABLE 9.2 (Continued)

Advantages and Disadvantages

The shilling attack detection technique	Advantages	Disadvantages
9. The semi-supervised learning using shilling attack detection (SEMI-SAD) (Wu et al. 2011).	1. This technique improves the naïve Bayes classifier by using expectation maximization (EM) to detect the user profiles. 2. Semi-SAD is very efficient and effective.	1. In this method classification of any big data is a real challenge. 2. Training needs a lot of computation time.
10. Support vector machine and target item analysis (SVM-TIA) (Zhou et al. 2016).	1. This method eases the class unbalance problem in classification.	1. In this method, potential attack profiles' set are analyzed, due to which some genuine profiles are misjudged as attack profiles.
11. Double M detector (DMD) and heterogeneous information network (HIN) method (Zhang et al. 2019).	1. DMD utilizes the interconnection between client–item and client–client using HIN. It enables the shilling attack identification for CFRSs. 2. This method works effectively in detecting the rating attacks as well as paying attention to the relation attacks. 3. DMD is also effective for hybrid attacks.	1. This system requires measurement of the relatedness of different types of objects for recommendation.
12. Detection based on statistical model (Wang et al. 2018)	1. A statistical model is idealized form of items rating generator process. 2. The probability of any event can be calculated by the statistical assumption which constitutes statistical model.	1. This model requires some information related to rating distribution of the data set. 2. The consumption of time gets affected as the scale of the data set increases. 3. Difficulty in handling leads to time bias and length-biased sampling.
13. Unsupervised method using the hidden Markov model and hierarchical clustering (UD-HMM) (Fuzhi et al. 2018).	1. Since this technique takes consideration of current state only, it does not require any prior knowledge. 2. It uses HMM, clusters genuine users and attackers based on ratings.	1. This method does not perform efficiently when detecting the collusive spammer on the sampled data set.

TABLE 9.2 (Continued)
Advantages and Disadvantages

The shilling attack detection technique	Advantages	Disadvantages
14. Attack detection in time series (Zhang et al 2006).	1. The time series of these sample average and sample entropy features can expose attack events by giving reasonable assumptions about their duration. 2. An optimal window size will be derived theoretically, only if the attack profile number is known. This is done to best identify the rating distribution changes, which is caused by attacks.	1. During attack, sometimes normal users' rating patterns also change, so it becomes difficult to identify shilling attackers.
15. Dynamic time interval segmentation technique (Xia et al. 2015).	1. This technique divides the life cycle of each item into several time intervals. This division is done dynamically at checkpoints. 2. It detects anomalous item through hypothesis test.	1. In this method, there exists an issue of improving the detection performance against target shifting attack. 2. It is not able to minimize the impact of attacks effectively, after identifying suspicious intervals of each item.
16. UnRap (Si and Li 2020).	1. This method is used to detect shilling profiles by analyzing user profile's rating deviation on target item.	1. It detects only the attack user profiles for individual times.
17. Data tracking (Qi et al. 2018)	1. Big Data processing is adapted in this technique. 2. Detection efficiency of the data tracking method is high.	1. Data tracking is not applicable to handle large data for distributed system.

9.7 Conclusion

Recommender system (RS) is an application that helps users to select relevant products from the internet. To reduce the damage of a shilling attack and maintain good quality of recommendation, the security of RSs is a significant issue. However, fake user profiles created by SA methods can be accurately detected by recent SA detection techniques. The ratting pattern of these SA techniques is different from real users. Being one of the most efficient ways for handling the problem of information overload, CFRSs are very much vulnerable to numerous shilling attacks due to insertion of variety of malicious user profiles in the system, which effects the user recommendations. In this chapter, at first the profile injection attack technique, collaborative filtering in recommender systems and shilling attack detection scheme have been discussed. Second, the shilling attack detection techniques have been classified based on the two parameters, namely, rating parameter and rating and time interval parameter on which they work. The shilling attack detection techniques have also been classified based on their output algorithms like attack profiles detection, attack items detection and attack profiles, and attack items detection. Then each of the shilling attack detection techniques have been discussed in brief. The techniques are analyzed and their advantages and disadvantages compared.

9.7.1 Future Direction

The future directions are working on the shortcomings of these shilling attack detection techniques. Like just in case of dynamic time interval segmentation technique, the detection performance against target shifting attack requires to be enhanced. A shortcoming of data tracking technique is that it is not applicable on distributed system to handle giant data. The problem with HHT-SVM is that, when new kinds of attacks are conducted, it becomes difficult to detect attack profiles. SVM classifier also requires being re-trained offline. Overcoming these problems are the future directions.

References

Bhaumik R, Burke R, Mobasher B (2007). Effectiveness of crawling attacks against web-based recommender systems. Proceedings of the 5th workshop on intelligent techniques for web personalization (ITWP-07).

Bhaumik R, Mobasher B, Burke R. (2011). A clustering approach to unsupervised attack detection in collaborative recommender systems. Proceedings of the International Conference on Data Mining (DMIN). The Steering Committee of the World Congress in Computer Science, Computer Engineering and Applied Computing (World Comp), p. 1.

Bhaumik R, Williams C, Mobasher B, Burke R (2006 Jul 16). Securing collaborative filtering against malicious attacks through anomaly detection. In Proceedings of the 4th Workshop on Intelligent Techniques for Web Personalization (ITWP'06), Boston. Vol. 6, p. 10.

Bilge A, Gunes I, Polat, H. (2014a) Robustness analysis of privacy-preserving model-based recommendation schemes. *Expert Syst Appl*, 41(8): 3671–3681.

Bilge A, Ozdemir Z, Polat H. (2014b). A novel shilling attack detection method. *Proc Computer Science*, 31: 165–174.

Bilge A, Polat H. (2013) A comparison of clustering-based privacy-preserving collaborative filtering schemes. *Appl Soft Comput*, 13(5): 2478–2489.

Burke R, Mobasher B, Williams C. (2006). Classification features for attack detection in collaborative recommender systems. International Conference on Knowledge Discovery and Data Mining, pp. 17–20.

Chirita P-A, Nejdl W, Zamfir C. (2005). Preventing shilling attacks in online recommender systems. In Proceedings of the 7th annual ACM international workshop on Web information and data management, pp. 67–74.

Deng, Z. J., Zhang, F., & Wang, S. P. (2016, July). Shilling attack detection in collaborative filtering recommender system by PCA detection and perturbation. In Proceedings of *International Conference on Wavelet Analysis and Pattern Recognition (ICWAPR)*, pp. 213–218.

Fuzhi Z, Zening Z, Peng Z (2018). UD-HMM: An unsupervised method for shilling attack detection based on hidden Markov model and hierarchical clustering. *Knowledge-Based Systems*, 148: 146–166.

Fuzhi Z, Zhou Q. (2014). HHT–SVM: An online method for detecting profile injection attacks in collaborative recommender systems. *Knowledge-Based Systems* 65: 96–105.

Gunes I, Kaleli C, Bilge A, et al. (2014). Shilling attacks against recommender systems: a comprehensive survey. *Artificial Intelligence Review*, 42(4): 767–799.

Jolliffe I. (2011). *Principal Component Analysis*. Springer, Berlin, pp. 1094–1096.

Karthikeyan P, Selvi ST, Neeraja G, Deepika R, Vincent A, Abinaya V. (2017, January). Prevention of shilling attack in recommender systems using discrete wavelet transform and support vector machine. In 2016 Eighth International Conference on Advanced Computing (ICoAC) IEEE, pp. 99–104.

Lam SK, Riedl J (2004). Shilling recommender systems for fun and profit. Proceedings of the 13th international conference on World Wide Web. ACM: pp. 393–402.

Mehta B (2007). Unsupervised shilling detection for collaborative filtering. AAAI. pp. 1402–1407.

Mehta B, Hofmann T, Fankhauser P. (2007) Lies and propaganda. Detecting spam users in collaborative filtering[C]. Proceedings of the 12th international conference on intelligent user interfaces. ACM: pp. 14–21.

Mehta B, Nejdl W (2009). Unsupervised strategies for shilling detection and robust collaborative filtering [J]. *User Modeling and User-Adapted Interaction*, 19(1–2): 65–97.

O'Mahony, Michael P. (2004). Towards robust and efficient automated collaborative filtering. PhD diss., University College Dublin.

Qi L, Huang H, Wang P and Wang R. (2018, August). Shilling Attack Detection Based on Data Tracking. In 2018 13th International Conference on Computer Science and Education (ICCSE), IEEE, pp. 1–4.

Sarwar, B., Karypis, G., Konstan, J., & Riedl, J. (2001, April). Item-based collaborative filtering recommendation algorithms. In Proceedings of the 10th international conference on World Wide Web: pp. 285–295.

Si, M., & Li, Q. (2020). Shilling attacks against collaborative recommender systems: a review. *Artificial Intelligence Review*, 53(1), 291–319.

Wei R, Shen H. (2016). An Improved Collaborative Filtering Recommendation Algorithm against Shilling Attacks. 17th International Conference on Parallel and Distributed Computing, Applications and Technologies, pp. 330–335.

Wang Y et al. (2018). A comparative study on shilling detection methods for trust-worthy recommendations. *Journal of Systems Science and Systems Engineering* 27(4): 458–478.

Wu Z, Cao J, Mao B, Wang Y. (2011, October). Semi-SAD: applying semi-supervised learning to shilling attack detection. In Proceedings of the fifth ACM conference on Recommender systems. ACM. pp. 289–292.

Xia H., Fang B, Gao M, Ma H, Tang Y, Wen J. (2015). A novel item anomaly detection approach against shilling attacks in collaborative recommendation systems using the dynamic time interval segmentation technique. *Information Sciences* 306: 150–165.

Xin, Z, Hong X, Yuqi S (2019). Meta-path and matrix factorization based shilling detection for collaborate filtering. International Conference on Collaborative Computing: Networking, Applications and Worksharing. Springer, Cham, pp. 3–16.

Yang, Z., et al. (2016). Re-scale AdaBoost for attack detection in collaborative filtering recommender systems. *Knowledge-Based Systems* 100: 74–88.

Zhang, S, Chakrabarti A, Ford J, Makedon F (2006). Attack detection in time series for recommender systems. In Proceedings of the 12th ACM SIGKDD international conference on Knowledge discovery and data mining, ACM. pp. 809–814.

Zhou W et al. (2016). SVM-TIA a shilling attack detection method based on SVM and target item analysis in recommender systems. *Neurocomputing* 210: pp. 197–205.

10

Machine Learning Applications in Real-World Time Series Problems

Antonio Manuel Durán-Rosal[1,*] and David Guijo-Rubio[2]

[1]Department of Quantitative Methods, Universidad Loyola Andalucía, Córdoba, Spain

[2]Department of Computer Science and Numerical Analysis, University of Córdoba, Córdoba, Spain

*amduran@uloyola.es

CONTENTS

10.1 Introduction

This first section introduces the topic presented and the related state-of-the-art developments. Time series data mining (TSDM) mainly consists of the following tasks: anomaly detection (Blázquez-García et al., 2020), classification (Ismail-Fawaz et al., 2019), analysis and preprocessing (Hamilton, 1994), segmentation (Keogh et al., 2004), clustering (Liao, 2005) and prediction (Weigend, 2018). More concretely, this chapter is focused on the applications of time series preprocessing, segmentation and prediction to real-world problems.

Time series analysis and preprocessing are considered previous steps for other TSDM tasks. One of the most important tasks is the imputation of missing values in time series, which is essential for the application of succeeding methods successfully. It is very common that real-time series, especially those collected by sensors, buoys or real physics systems, can have gaps of information given that they may stop working due to different reasons. Within this context, many applications have been proposed in recent years. Typically, the imputation of missing values has been done by interpolating the observational feature space, without considering any hidden dynamics. Zhou and Huang (2018) proposed a model that would capture the latent complex with a novel iterative imputing network and demonstrated its performance by its application to a meteorological benchmark data set outperforming the state-of-the-art methods. More recently and following the idea of capturing the dynamics from time series, Belda et al. (2020) have developed a software called DATimeS, which uses machine learning fitting algorithms such as Gaussian process regression (GPR). It has resulted in powerful software for the reconstruction of image vegetation time series, which tend to be discontinuous because of cloud cover.

Time series segmentation consists in the division of a time series into several consecutive subsequences, trying to achieve different objectives. More formally, they are defined as follows: given a time series $Y = \{y_i\}$, for the time indexes $i = 1...N$, the procedure attempts to separate the values of Y into a set of consecutive l segments. In this way, the time indexes are split into $S_1 = \{y_1,...,y_{t1}\}$, $S_2 = \{y_{t1},...,y_{t2}\}$, ..., $S_l = \{y_{t_{l-1}},...,y_N\}$, where the ascending ordered cut points are expressed as $t_1 < t_2 < \cdots < t_{l-1}$. Considering the

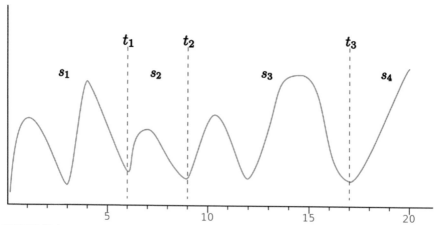

FIGURE 10.1

Segmentation procedure for a time series of length $N = 20$ and four segments. $S_1 = \{y_1, y_2, y_3, y_4, y_5, y_6\}$, $S_2 = \{y_6, y_7, y_8, y_9\}$, $S_3 = \{y_9, y_{10}, y_{11}, y_{12}, y_{13}, y_{14}, y_{15}, y_{16}, y_{17}\}$ and $S_4 = \{y_{17}, y_{18}, y_{19}, y_{20}\}$, resulting from cut points $t_1 = 6, t_2 = 9$ and $t_3 = 17$.

transition from one segment to the next, each cut point belongs to two segments, in such a way that they belong to the previous and to the next segment. Figure 10.1 graphically represents a segmentation procedure.

As abovementioned, segmentation is used to achieve different objectives. In the literature, there are two main groups of objectives. On the one hand, this technique is carried out to discover segment similarities or useful patterns over time. Methods within this context try to optimize the division of time series and then group the segments into different clusters. Several methods have been proposed during the last two decades. Initially, Abonyi et al. (2003) stated that all points in a time series belonging to the same cluster are contiguous in time. After that, many authors proposed methods to group segments instead of points. Tseng et al. (2009) developed an algorithm in which similarities in wavelet space are considered to guide the segmentation, resulting in a clustering of segments of different length. Besides, a significant clustering of subsequent time series was addressed using two efficient methods (Rakthanmanon et al., 2012). As can be seen, all these segmentation procedures require clustering algorithms, given that they aim to make groups of segments in order to discover useful similarities. In this sense, several algorithms for time series clustering have been proposed recently (Guijo-Rubio et al., 2020a), aiming to obtain time series groups with similar characteristics or based on segments typologies.

On the other hand, time series segmentation is also applied for approximating time series; in other words, to reduce the number of time series points. In this case, the procedure is performed by selecting those points whose approximation is the most accurate regarding the real-time series, that is, the approximation aims to minimize the information loss or the approximation error. One of the main purposes is to mitigate the difficulty of processing and memory requirements. A well-known approach in this context is the use of piecewise linear approximations, where linear regressions or interpolations are used for modelling each segment (Keogh et al., 2004). Moreover, Fu (2011) presented an approach for approximating time series by defining it as an optimization problem that could be solved by evolutionary algorithms. Lately, other authors have developed a novel approach based on connected lines under a predefined maximum error bound (Zhao et al., 2016).

Finally, the prediction of a time series is easily and formally defined as follows: given a time series $Y = \{y_i\}(i = 1 \ldots N)$, the prediction consists in the determination of the value y_{n+t}, T being a future instant of time. This procedure learns from the known past values in order to generate a model able to estimate future ones accurately. Traditional statistical models are still widely used for this task, for example, the COVID-19 pandemic spread in Saudi Arabia has been forecasted by AutoRegressive Integrated Moving Average (ARIMA) models (Alzahrani et al., 2020). Nevertheless, nowadays there has been an increasing interest in developing novel ML algorithms, such as artificial neural networks (ANNs), or even more advanced approaches, such as evolutionary ANNs (EANNs) or long short-term memory (LSTM), which have been applied to finances (Chandra and Chand, 2016) and to weather prediction (Karevan and Suyken, 2020), respectively.

The following section presents a set of real-world applications tackled by the authors of this chapter.

10.2 Real-World Applications of TSDM Using ML algorithms

10.2.1 Massive Missing Data Reconstruction in Wave Height Time Series

Oceanography buoys are instruments used to measure different wave properties. Their accuracy and availability are crucial for several essential operations, like coastal structures maintenance, safe ship navigation or wave height converters, among others (López et al., 2013). Although there are many agencies offering support or maintaining the buoys, as is the case of the National Data Buoy Centre (NDBC) in the United States, many unexpected events make buoys to break down (e.g. prolonged storms, harmful accidents, long maintenance periods and so on). In this way, it causes data gaps until the buoy is completely repaired. Even though some methods can be applied even with the presence of missing data in the time series, the number of which need complete data is much higher (Thomson and Emery, 2014). For this reason, the reconstruction of missing values in time series has attracted a great interest in this field.

In the last two decades, there has been an exponential increase in the number of works applying ML techniques to data recovery. One of the first works in the area was done by Bhattacharya et al. (2003), in which ANNs are used to retrieve missing values of wave height data in the North Sea. Then, feedforward multilayer perceptrons (MLPs) and recurrent neural networks (RNN), trained by conjugate gradient, were presented as accurate methods for reconstructing the wave height in Turkey (Balas et al., 2004). Besides, a back-propagation algorithm for training ANNs was also used in the Atlantic area with the same purpose of recovering data from buoys (Gunaydin and Panchang, 2008). More works used ANNs and novel variants during these years, such as ANNs combined with *K*-nearest neighbour algorithm (Zamani et al., 2008) or ANNs trained by the rough set theory (Setiawan et al., 2008), among others.

Furthermore, ANNs have also been combined with some other methodologies such as the work of Londhe (2008) where ANNs are combined with genetic programming to estimate missing wave height using neighbours' stations. In Mahjoobi et al. (2008), a neuro-fuzzy inference system was successfully applied over Lake Ontario. And finally, physical methods have also been combined with ML approaches for wave height estimation (Casa-Prat et al., 2014).

As most ML models lack interpretability inherent to these techniques, we proposed a new method in Durán-Rosal et al. (2015). In this paper, a product unit neural network trained by an evolutionary algorithm (EPUNN) was proposed through a two-staged procedure. This methodology was applied to six buoys located at the Gulf of Alaska, United States.

The first stage consisted in performing an initial reconstruction using transfer functions and neighbour correlation method. On the one hand, transfer functions are based on the analysis of the correlation and the estimation of a gappy time series by means of a complete one (note that the correlation between them needs to be higher than a predefined threshold). On the other hand, the neighbour correlation method estimates these missing values by adding information from the highly correlated buoys.

Once both methods have recovered the missing values, we keep the best one, that is, the one resulting in the smallest error. After that, the best recovery for each time series is used as input for the EPUNN of the second stage. In this sense, the two most correlated inputs (concerning the original time series) are used to get the definitive reconstructed time series.

The results achieved for the six buoys located at the Gulf of Alaska indicated that EPUNNs are suitable for this kind of problems, given their accuracy and their interpretability. Besides, they can be represented as linear models by applying natural logarithm to the inputs for easing the understanding and interpretation. As a remark, better reconstructions are achieved for coastal buoys, given that the availability of values was higher.

10.2.2 Detection and Prediction of Tipping Points

Palaeoclimatology is a popular field of palaeogeography science; it consists of the study of Earth's climate characteristics, and concretely, it aims to provide the best possible description of the climate for a certain period of the history. Moreover, palaeoclimatology analyses significant climate variations and the reasons behind them. Some points of no return, thresholds and phase changes are widespread in nature and often nonlinear. These could have a high impact on the Earth's climate being difficult to anticipate (Wassmann et Lenton, 2012).

Focusing on this problem, several researchers on dynamic systems, such as climate systems, claim that they present critical transitions. These transitions are commonly called tipping points (TPs), also known as 'small things that can make a big difference'. When a small change in a climate system causes a strong non-linear response in its internal dynamics, leading to a change in its future state, it is a climate TP. In this sense, the main climatic changes on Earth are determined by the transition points in the temperature time series, or one of their proxies such as the concentration of oxygen isotopes in the glaciers, which makes them go from one stable state to another.

Early detection of TPs by analysing their causes and what happens before them has been considered of significant impact since they can affect millions of lives. Lenton (2011) proposed the differentiation of many types of TPs and presented some indicators that may help to detect them, like increased auto-correlation of series values. Besides, Dakos et al. (2012) introduced more par-ticular techniques related to data processing and indicators. They studied several methods using simulated ecological data and arrived at principal conclusions: there is no single best indicator to detect a transition and that all methods require specific data processing. The previous works have two main disadvantages: the need of intensive preprocessing of the data and the appli-cation of a specific treatment for each type of TP. In this regard, we proposed two works.

In the first work, a genetic algorithm (GA), in combination with an unsupervised ML technique, was proposed as a segmentation algorithm to correctly identify the group of segments, where some of them represented TP transitions (Nikolaou et al., 2015). Based on a random division, the algorithm optimizes the cut points aiming to improve the quality of the clustering, according to their statistical similarities (variance, asymmetry, kurtosis, slope, approximation error and autocorrelation).

This methodology was tested not only using synthetic data sets generated from recognized dynamic systems but also on two real-world paleoclimate oxygen isotope datasets: GISP2 (Greenland Ice Sheet Project Two) and NGRIP (North Greenland Ice Core Project) δ^{18} O time series with a resolution of 20 years and an average of 5 points to reduce short fluctuation. The experiments resulted in an effective detection and dis-covery of similarities and differences in the dynamic characterization of TPs, called Dangaard–Oeschger (DO) events in these time series. The algo-rithm determined that the increase in autocorrelation, variance and mean squared error can be considered as warning signals for almost all TPs. Moreover, the proposed approach provided a novel visualization tool for climate time series analysis.

In the second work, we introduced some improvements and developed a prediction model derived from the obtained segments (Pérez-Ortiz et al., 2019). Once the segmentation stage finished, its evaluation was automatic-ally done by performing a comparison with an ideal segmentation carried out by experts in the field and by measuring the algorithm stability. The algorithm was also supplied with 10 metrics for the clustering validation, in order to increase the robustness in the search for the best segmentation, thus resulting in a better performance in terms of TPs detection. For the prediction phase, the segments obtained by the GA were then transformed into a binary problem, with the goal of determining whether the next state represented a DO event.

The results achieved confirmed that the improvements made outperformed those of the previous algorithm detecting all but two TPs in both data sets. The experiments also demonstrated that the proposed diagnostic-predictive ML-based model, more specifically a decision tree, was feasible and generated very promising results. This behaviour was produced given that the system's prediction-monitoring period lasts several decades, and that the time series already contained in its record abrupt climate changes of the same type. The above-mentioned aspects open an interesting field for future research.

10.2.3 Prediction of Fog Formation

It is widely known that aviation is one of the means of transport most affected by adverse weather conditions. There are a lot of phenomena reducing the visibility, such as sandstorms or torrential rains. However, without any doubt, the most important and most common is fog formation. Fog is a meteorological phenomenon consisting in the suspension of almost microscopic particles of water droplets in the air, causing the reduction of horizontal visibility on the Earth's surface. It has been addressed from different points of view (Román-Gascón, 2015).

Foggy weather may result in air traffic flow disruptions, flight delays or even accidents, among others. More specifically, it has a major impact on the safety and efficiency of airport operations such as taxiing, take-off and landing operations (Bergot et al., 2007). Intending to ensure an enough efficiency, airport workers need to have the most reliable information, with the aim of anticipating these phenomena. Some studies have proposed developments of trustworthy systems to detect and predict fog formation.

Perhaps one of the most widespread methods of predicting fog formation is the use of numerical weather prediction (NWP) models (Román-Gascón et al., 2016). Nevertheless, accurate NWP models are computationally expensive. In recent years, ML methods have been successfully applied to fog prediction problems. Fabbian et al. (2007) proposed an ANN to predict fog events on an 18-hour schedule at Canberra International Airport in Australia. Similarly, Marzban et al. (2007) used ANNs to combine three sources of information to forecast ceiling and visibility at 39 airports in the northwestern United States. In the same year, Bremnes and Michaelides (2007) applied another ANN approach to fog formation prediction with a time horizon of up to 6 h in Norway. Then, a Bayesian model was successfully introduced in a fog prediction problem with a time horizon of 12 h in the Pacific Northwest of the United States (Chmielecky and Raftery, 2011). More recently, Colabone et al. (2015) proposed an MLP that uses meteorological data to help in the planning of flight activities of the Academia da Força Aérea (AFA). In Cornejo-Bueno et al. (2017), different ML regression techniques have been applied to fog prediction in Spain. These works tackled the problem of fog formation as regression approaches.

Although there are many sensors collecting meteorological data at airports, such as wind speed and direction, temperature, humidity and pressure, among others, in order to support the airport operations, the runway visual range (RVR) is the most important. RVR is a weather variable defined as the range over which the pilot of an aircraft on the centreline of a runway can see the markings on the runway surface or the lights that outline the runway or identify its centreline.

Taking into account the previous works and the RVR variable, we proposed two works in the sense that fog prediction is treated as class intervals, resulting in a binary and an ordinal problem, respectively. Both works are located at the airport of Valladolid since its location is a key point in the formation of fog during the cold months.

The first work is focused on the prediction of the RVR variable using data collected from sensors with a 6-hour time horizon prediction (Durán-Rosal et al., 2018). Fog events are considered when RVR is less than 1990 m. Thus, the problem becomes a binary classification in which the output is 1 if there is a fog event, and 0 otherwise. Given the unbalanced nature of the data set, that is, the number of fog events was much lower than clear events, a multi-objective evolutionary algorithm (MOEA) was developed to train the structure and weights of ANNs, using PUs, sigmoid units (SUs) and radial base functions (RBFs) as basis functions. The algorithm tried to optimize the overall accuracy and minimum sensitivity to address the problem of unbalancing.

The best model obtained by MOEA was an ANN with a PU basis function. This combination achieved a global accuracy of 82.24 per cent and the percentage of fog events correctly classified was 84.83 per cent. The resulting ANN had only two neurons in the hidden layer, showing the high interpretability of the model (see Figure 10.2), which was able to be explained in terms of physical properties of fog formation.

In the second work (Guijo-Rubio et al., 2018), three ordered classes were considered, increasing the difficulty of the problem. The consideration of these three classes was due to giving more precise information as a decision support system: FOG (RVR in the range [0, 1000] m), MIST (RVR in the range [1000, 1900] m) and CLEAR (RVR higher or equal to 1990 m). Besides, the prediction was carried out using a 24-hour schedule.

The work was based on the philosophy of autoregressive (AR) models, so the model needed some previous values for the prediction of a given one. To determine this number of previous values, three different sorts of window were proposed: fixed window (FW), dynamic window (DW) based on label change (DWLC) and DW based on variance change (DWVC). Since the problem was addressed under the ordinal paradigm, the following algorithms were proposed to solve it: proportional odds model (POM), support vector machines (SVM), support vector ordinal regression considering explicit constraints (SVOREX), support vector ordinal regression considering implicit constraints (SVORIM) and kernel discriminant learning for ordinal regression (KDLOR).

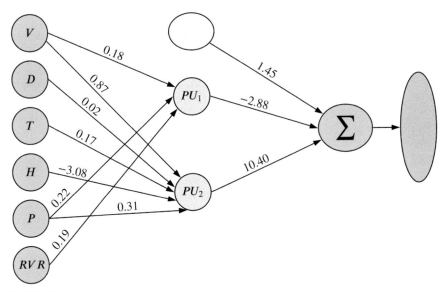

FIGURE 10.2
Best ANN obtained by MOEA using PUs as basis functions. *V* and *D* are the velocity and wind direction, respectively, *T* corresponds to the temperature, *H* represents the humidity, *P* is the pressure, and *RVR* is the runway visual range previously defined.

The results showed that the combination of a FW and a DW with the algorithm KDLOR is outstanding, producing the best results in terms of minimum sensitivity and AMAE (the average of ordinal classification error made for each class). This methodology could lead to an improvement in the safety and profitability of aviation operations at airports affected by low-visibility events.

10.2.4   Prediction of Other Convective Situations

As mentioned in Section 10.2.3., there is a great number of phenomena interrupting the proper operation of airports. Not only do they cause endless queues to the passengers but they also increase the workload of the operators and workers (Cao and Cai, 2016). Thus, anticipating this sort of phenomena has a significant impact nowadays. In this way, several proposals have been presented in the literature, trying to improve our knowledge about convective situations and aiming also to improve the current systems (Wang et al., 2014). At the present time, most of the approaches are based on stability indices derived from the temperature and humidity, such as the lifted index or the convective available potential energy index. Even though these indices have been proved to be useful in the prediction of conditions related with convective situations, their performance can be increased by combining them with ML techniques, which have also been proved to be excellent when dealing with atmospheric issues (McGovern et al., 2017).

A convective cloud formation problem was considered by the author of the chapter. Data collected from the airport of Madrid-Barajas (Spain) for five years (2011 to 2015) was used to present two novel approaches to the literature. The inputs considered in the problem were a combination of atmospheric variables measured over the vertical with some popular stability indices. The main goal of the study was the prediction of convective situations at 12 h time horizon. In this regard, two main proposals were made to this problem: the first one (Guijo-Rubio et al., 2020b) consisted in tackling the problem from the nominal classification point of view. Each event was given one of the following categories: clear (no convective clouds sighted), TCU (cumulus congestus clouds sighted), CB (cumulonimbus clouds sighted) and TS (thunderstop sighted). As can be imagined, this problem is highly unbalanced given that there were more clear patterns than TCU, CB or TS. Therefore, in this work, we analysed the performance of ANN trained by a MOEA, aiming to achieve the best global performance without prejudicing the smallest class (in this case, the TS). The results achieved demonstrated that hybrid ANN (mixing basis functions in the hidden layer) are an excellent approach for this kind of imbalanced problems.

On the other hand, the second proposal made in this field was tackled from the ordinal point of view (Guijo-Rubio et al., 2020c). Besides considering that the classes followed an order relationship among them, in this work, an oversampling method is previously applied as a preprocessing step, aiming to balance the data set as much as possible. After that, ordinal methods were applied to the resulting data set. SVORIM achieved the best results, improving also the results achieved by terminal aerodrome forecasts (TAFs), which are products available on the aerodrome that give concise statements of the expected meteorological conditions, for almost all the categories.

10.3   Summary and Other Related Works

This chapter presents a set of TSDM applications using ML. The tasks the chapter focuses on are reconstruction, segmentation and prediction, all of them applied to time series data. The main applications presented, from the reconstruction of missing values in wave height time series to the prediction of low-visibility situations in airports, provide the data in the form of time series. The combination of time series data with the application of automatic ML techniques to solve these real-world problems has been proved to have an excellent performance.

The authors of this chapter have addressed some other applications in real-world problems since time series can be found in a wide variety of fields, as was mentioned above. A GA was proposed in combination with a likelihood-based segmentation procedure to automatically recognize financial patterns in European stock market indexes (Durán-Rosal et al., 2017a). The detection and prediction of extreme waves using the retrieved time series obtained by oceanographic buoys were also solved by applying a two-stage algorithm (Durán-Rosal et al., 2017b). First, a segmentation algorithm, resulting from an evolutionary approach hybridized with a likelihood-based segmentation involving a beta distribution, was used to detect the extreme events. Second, an evolutionary ANN was proposed to predict them. Finally, in Guijo-Rubio et al. (2020d), evolutionary ANNs were successfully applied for the prediction of global solar radiation at the radiometric station of Toledo (Spain) using satellite-based measurements.

Apart from this, there is a vast amount of works using time series in other applications. Therefore, the potential of this type of temporal data is significant due to the wide variety of solutions given to these problems by using ML techniques. The future lies in the massive collection of data in real time, its processing and the instantaneous generation of automatic ML techniques to model and extract knowledge from them.

References

Abonyi, J., Feil, B., Nemeth, S., Arva. P. 2003. Fuzzy clustering based segmentation of time-series. In *Advances in Intelligent Data Analysis V*, 275–285. https://doi.org/10.1007/978-3-540-45231-7_26

Alzahrani, S. I., Aljamaan, I. A., Al-Fakih, E. A. 2020. Forecasting the spread of the COVID-19 pandemic in Saudi Arabia using ARIMA prediction model under current public health interventions. *Journal of Infection and Public Health, 1*(7), 914–919. https://doi.org/10.1016/j.jiph.2020.06.001

Belda, S., Pipia, L., Morcillo-Pallarés, P., Rivera-Caicedo, J. P., Amin, E., De Grave, C., Verrelst, J. 2020. DATimeS: A machine learning time series GUI toolbox for gap-filling and vegetation phenology trends detection. *Enviromental Modelling and Software 127*, 104666. https://doi.org/10.1016/j.envsoft.2020.104666

Balas, C., Koç, L., Balas, L. 2004. Predictions of missing wave data by recurrent neuronets. *Journal of Waterway, Port, Coastal, and Ocean Engineering, 130*(5), 256–265. https://doi.org/10.1061/(ASCE)0733-950X(2004)130:5(256)

Bergot, T., Terradellas, E., Cuxart, J., Mira, A., Liechti, O., Mueller, M., Nielsen, N.W. 2007. Intercomparison of single-column numerical models for the prediction of radiation fog. *Journal of Applied Meteorology and Climatology, 46*, 504–521. https://doi.org/10.1175/JAM2475.1

Bhattacharya, B., Shrestha, D., Solomatine, D. 2003. Neural networks in reconstructing missing wave data in sedimentation modelling. In *Proceedings of the 30th IAHR Congress, 500*, 770–778.

Blázquez-García, A., Conde, A., Mori, U., Lozano, J. A. 2020. A review on outlier/anomaly detection in time series data. *ArXiv:2002.04236 [Cs]*, https://arxiv.org/abs/2002.04236

Bremnes, J. B., Michaelides, S. C. 2007. Probabilistic visibility forecasting using neural networks. In *Fog and Boundary Layer Clouds: Fog Visibility and Forecasting*. Springer, pp. 1365–1381. https://doi.org/10.1007/s00024-007-0223-6

Cao, Z., Cai, H. 2016. Identification of forcing mechanisms of convective initiation over mountains through high-resolution numerical simulations. *Advances in Atmospheric Sciences, 33*(10), 1104. https://doi.org/10.1007/s00376-016-6198-4

Casas-Prat, M., Wang, X. L., Sierra, J. P. 2014. A physical-based statistical method for modeling ocean wave heights. *Ocean Modelling, 73*, 59–75. https://doi.org/10.1016/j.ocemod.2013.10.008

Chandra, R., Chand, S. 2016. Evaluation of co-evolutionary neural network architectures for time series prediction with mobile application in finance. *Applied Soft Computing, 49*, 462–473. https://doi.org/10.1016/j.asoc.2016.08.029

Chmielecki, R. M., Raftery, A. E. 2011. Probabilistic visibility forecasting using Bayesian model averaging. *Monthly Weather Review, 139*, 1626–1636. https://doi.org/10.1175/2010MWR3516.1

Colabone, R. D. O., Ferrari, A. L., Vecchia, F. A. D. S., Tech, A. R. B. 2015. Application of artificial neural networks for fog forecast. *Journal of Aerospace Technology and Management, 7*, 240–246. https://doi.org/10.5028/jatm.v7i2.446

Cornejo-Bueno, L., Casanova-Mateo, C., Sanz-Justo, J., Cerro-Prada, E., Salcedo-Sanz, S. 2017. Efficient prediction of low-visibility events at airports using machine-learning regression. *Boundary-Layer Meteorology, 165*, 349–370. https://doi.org/10.1007/s10546-017-0276-8

Dakos, V., Carpenter, S. R., Brock, W. A., Ellison, A. M., Guttal, V., Ives, A. R., Kefi, S., Livina, V., Seekell, D. A., Van Nes, E. H, et al. 2012. Methods for detecting early warnings of critical transitions in time series illustrated using simulated ecological data. *PloS One, 7*(7), e41010. https://doi.org/10.1371/journal.pone.0041010

Durán-Rosal, A. M., Hervás-Martínez, C., Tallón-Ballesteros, A. J., Martínez-Estudillo, A. C., Salcedo-Sanz, S. 2015. Massive missing data reconstruction in ocean buoys with evolutionary product unit neural networks. *Ocean Engineering, 117*, 292–301. https://doi.org/10.1016/j.oceaneng.2016.03.053

Durán-Rosal, A. M., de la Paz-Marín, M., Gutiérrez, P. A., Hervás-Martínez, C. 2017a. Identifying market behaviours using European Stock Index time series by a hybrid segmentation algorithm. *Neural Processing Letters, 46*(3), 767–790. https://doi.org/10.1007/s11063-017-9592-8

Durán-Rosal, A. M., Fernández, J. C., Gutiérrez, P. A., Hervás-Martínez, C. 2017b. Detection and prediction of segments containing extreme significant wave heights. *Ocean Engineering, 142*, 268–279. https://doi.org/10.1016/j.oceaneng.2017.07.009.

Durán-Rosal, A. M., Fernández, J. C., Casanova-Mateo, C., Sanz-Justo, J., Salcedo-Sanz, S., Hervás-Martínez, C. 2018. Efficient fog prediction with multi-objective evolutionary neural networks. *Applied Soft Computing 70*, 347–358. https://doi.org/10.1016/j.asoc.2018.05.035.

Fabbian, D., de Dear, R., Lellyett, S., 2007. Application of artificial neural network forecasts to predict fog at Canberra international airport. *Weather Forecast, 22,* 372–381. https://doi.org/10.1175/WAF980.1.

Fu, T. C. (2011). A review on time series data mining. *Engineering Applications of Artificial Intelligence,* 24(1), 164–181.

Guijo-Rubio, D., Gutiérrez, P. A., Casanova-Mateo, C., Sanz-Justo, J., Salcedo-Sanz, S., Hervás-Martínez, C. 2018. Prediction of low-visibility events due to fog using ordinal classification. *Atmospheric Research, 214,* 64–73. https://doi.org/ 10.1016/j.atmosres.2018.07.017.

Guijo-Rubio, D., Durán-Rosal, A. M., Gutiérrez, P. A., Troncoso, A., Hervás-Martínez, C. 2020a. Time-series clustering based on the characterization of segment typologies. *IEEE Transactions on Cybernetics* (early access). http://doi.org/10.1109/ TCYB.2019.2962584

Guijo-Rubio, D., Gutiérrez, P. A., Casanova-Mateo, C., et al. 2020b. Prediction of convective clouds formation using evolutionary neural computation techniques. *Neural Computing and Applications, 33,* 13917–13929. https://doi.org/10.1007/ s00521-020-04795-w

Guijo-Rubio, D., Casanova-Mateo, C., Sanz-Justo, J., Gutiérrez, P. A., Cornejo-Bueno, S., Hervás, C., Salcedo-Sanz, S. 2020c. Ordinal regression algorithms for the analysis of convective situations over Madrid-Barajas airport. *Atmospheric Research, 236,* 104798. https://doi.org/10.1016/j.atmosres.2019.104798

Guijo-Rubio, D., Durán-Rosal, A. M., Gutiérrez, P. A., et al. 2020d. Evolutionary artificial neural networks for accurate solar radiation prediction. *Energy, 210,* 118374. https://doi.org/10.1016/j.energy.2020.118374

Gunaydin, K. 2008. The estimation of monthly mean significant wave heights by using artificial neural network and regression methods. *Ocean Engineering, 35*(14–15), 406–1415. https://doi.org/10.1016/j.oceaneng.2008.07.008

Karevan, Z., & Suykens, J. A. (2020). Transductive LSTM for time-series prediction: An application to weather forecasting. Neural Networks, 125, 1–9.

Hamilton, J. D. 1994. *Time Series Analysis,* volume 2. Princeton University Press, Princeton, NJ.

Ismail Fawaz, H., Forestier, G., Weber, J. et al. 2019. Deep learning for time series classification: a review. *Data Mining and Knowledge Discovery 33,* 917–963. https:// doi.org/10.1007/s10618-019-00619-1

Keogh, E., Chu, S., Hart, D., Pazzani, M. 2004. Segmenting time series: A survey and novel approach. *Data mining in Time Series Databases,* 1–21. https://doi.org/ 10.1142/9789812565402_0001

Lenton, T. M. 2011. Early warning of climate tipping points. *Nature Climate Change, 1,* 201–209. https://doi.org/10.1038/nclimate1143

Liao, T. W. 2005. Clustering of time series data—a survey. *Pattern Recognition, 38*(11), 1857–1874. https://doi.org/10.1016/j.patcog.2005.01.025

Londhe, S. 2008. Soft computing approach for real-time estimation of missing wave heights. *Ocean Engineering, 35*(11–12), 1080–1089. https://doi.org/10.1016/ j.oceaneng.2008.05.003

López, I., Andreu, J., Ceballos, S., de Alegría, I. M., Kortabarria, I. 2013. Review of wave energy technologies and the necessary power-equipment. *Renewable and Sustainable Energy Reviews, 27,* 413–434. https://doi.org/10.1016/ j.rser.2013.07.009

Mahjoobi, J., Etemad-Shahidi, A., Kazeminezhad, M. 2008. Hindcasting of wave parameters using different soft computing methods. *Applied Ocean Research,* 30(1), 28–36. https://doi.org/10.1016/j.apor.2008.03.002

McGovern, A., Elmore, K. L., Gagne, D. J., Haupt, S. E., Karstens, C. D., Lagerquist, R. 2017. Using artificial intelligence to improve real-time decision-making for high-impact weather. *Bulletin of the American Meteorological Society,* 98(10), 2073–2090. https://doi.org/10.1175/BAMS-D-16-0123.1

Marzban, C., Leyton, S., Colman, B. 2007. Ceiling and visibility forecasts via neural networks. *Weather and Forecasting,* 22, 466–479. https://doi.org/10.1175/WAF994.1

Nikolaou, A., Gutiérrez, P. A., Durán, A., Dicaire, I., Fernández-Navarro, F., & Hervás-Martínez, C. (2015). Detection of early warning signals in paleoclimate data using a genetic time series segmentation algorithm. *Climate Dynamics,* 44(7), 1919–1933.

Pérez-Ortiz, M., Durán-Rosal, A. M., Gutiérrez, P. A., Sánchez-Monedero, J., Nikolauou, A., Fernández-Navarro, F., Hervás-Martínez, C. 2019. On the use of evolutionary time series analysis for segmenting paleoclimate data. *Neurocomputing, 326–327,* 3–14. https://doi.org/10.1016/j.neucom.2016.11.101

Rakthanmanon, T., Keogh, E. J., Lonardi, S., Evans, S. 2012. MDL-based time series clustering. *Knowledge and Information Systems,* 33(2), 371–399. https://doi.org/10.1007/s10115-012-0508-7

Román-Gascón, C. 2015. Radiation fog, gravity waves and their interactions with turbulence in the atmospheric boundary layer. Ph.D. thesis, Universidad Complutense de Madrid.

Román-Gascón, C., Steeneveld, G., Yagüe, C., Sastre, M., Arrillaga, J., Maqueda, G. 2016. Forecasting radiation fog at climatologically contrasting sites: evaluation of statistical methods and WRF. *Quarterly Journal of the Royal Meteorological Society,* 142(695), 1048–1063. https://doi.org/10.1002/qj.2708

Setiawan, N., Venkatachalam, P., Hani, A. 2008. Missing attribute value prediction based on artificial neural network and rough set theory. International Conference on Biomedical Engineering and Informatics (MEI2008), 1, 306–310. https://doi.org/10.1109/BMEI.2008.322

Thomson, R. E., Emery, W. J. 2014. *Data Analysis Methods in Physical Oceanography.* Elsevier, San Diego, CA.

Tseng, V. S., Chen, C. H., Huang, P. C., Hong, T. P. 2009. Cluster-based genetic segmentation of time series with DWT. *Pattern Recognition Letters,* 30(13), 1190–1197. https://doi.org/10.1016/j.patrec.2009.05.013

Wang, H., Luo, Y., Jou, B. J. D. 2014. Initiation, maintenance, and properties of convection in an extreme rainfall event during SCMREX: Observational analysis. *Journal of Geophysical Research: Atmospheres, 199*(23), 13–206. https://doi.org/10.1002/2014JD022339

Wassmann, P., Lenton, T. M. 2012. Arctic tipping points in an earth system perspective. *AMBIO, 41,* 1–9. https://doi.org/10.1007/s13280-011-0230-9

Weigend, A. S. 2018. *Time Series Prediction: Forecasting the Future and Understanding the Past.* Routledge, London.

Zamani, A., Solomatine, D., Azimian, A., Heemink, A. 2008. Learning from data for wind-wave forecasting. *Ocean Engineering,* 35(10), 953–962. https://doi.org/10.1016/j.oceaneng.2008.03.007

Zhao, H., Dong, Z., Li, T., Wang, X., Pang, C. 2016. Segmenting times series with connected lines under maximum error bound. *Information Sciences, 345,* 1–8. https://doi.org/10.1016/j.ins.2015.09.017

Zhou, J., Huang, Z. 2018. Recover missing sensor data with iterative imputing network. In Workshops at the Thirty-Second AAAI Conference on Artificial Intelligence.

11

Prediction of Selective Laser Sintering Part Quality Using Deep Learning

Lokesh Kumar Saxena[1,*] and Pramod Kumar Jain[2]

[1]*Mechanical Engineering Department, J.M.I., New Delhi, India*

[2]*Department of Mechanical and Industrial Engineering, Indian Institute of Technology, Roorkee, India 247667*

Corresponding author. email:lokeshkrsax@rediffmail.com, lokeshkrsaxiitr@gmail.com

CONTENTS

11.1 Introduction

Manufacturing is the pillar of the economy that transforms raw materials into products. Manufacturing is faced with a growing need for individualisation,

DOI: 10.1201/9781003104858-11

quality, flexibility and efficiency. Numerous countries had shown a keen interest in strategic plans to speed up the faster transformation from existing mechanised, semi-automatic or automatic manufacturing to smart manufacturing, for example, the United States with Advanced Manufacturing Partnership and Germany with Industry 4.0. Smart manufacturing has not been formally defined yet. But its core concepts have been recognised by the researchers, namely, to monitor, control, integrate and optimise the manufacturing processes employing the elements such as cyber-physical systems (CPS), Internet of Things (IoT), big data, advanced data analytics and imparting intelligence to machines.[1–4]

Traditionally, manufacturing, especially subtractive manufacturing, removes the material from the surface of raw stock to make a product. In contrast, additive manufacturing (AM) is an important modern digital manufacturing industrial paradigm. It is eliciting a keen interest in manufacturing.[5–14] It produces the product by adding the material layer by layer employing three-dimensional (3D) models. These models are made using computer-aided design software (CAD). AM ensures various benefits such as producing complex shaped products with optimised structures in topology and difficult to produce by conventional casting/forging, producing the novel material properties, for example, dislocation networks[15] and decreasing the material waste to save the cost for industry. But AM products also possess unique defects such as anisotropic microstructure in the parallel and perpendicular and directions with respect to the printing direction; porosity owing to gas entrapment and imperfect fusion; and distortion owing to residual stress resulting from faster temperature gradient and very steep cooling rate.[16] Therefore, it is important to determine the relationship between metallurgical properties of powder, AM process and the mechanical properties and the microstructure of AM products. There are various important parameters in the AM process responsible for the characteristics of end products. For instance, the processing parameters for selective laser sintering (SLS) are surrounding working temperature, laser scanning speed, layer thickness, scanning mode, hatch distance, laser power and interval time. The impacts of these parameters vary considerably with respect to the quality of the produced products. But it is very difficult to know the relationship between SLS parameters and product quality. The SLS process needs no support structure to produce parts. It can process various materials, for example, ceramics, nylon, polycarbonate, wax, nylon/glass, metal–polymer powders and composite.[17] The dimensional accuracy of SLS products is poor as compared to the conventional machining processes.[18] Thus, it is important to enhance the accuracy of SLS process. Li et al. studied the shrinkage parameters for SLS process.[19] John and Carl examined heat transfer, the energy delivery and sintering process.[20] Yang et al. employed the Taguchi method to examine the shrinkage compensation.[21] Masood et al. examined the orientation of products.[22] Bai et al. examined the temperature field for polymer–molybdenum powder.[23] Arni and Gupta analysed perpendicularity, parallelism and flatness tolerance.[24] Armillotta and Biggioggero studied the impact of part orientation and layer thickness on surface finish.[25] Shi et al. applied expert system with the neural network.[26] Product inaccuracy results due to material contraction in the SLS.[27] Contraction produces stress inside and deforms the parts. Power material properties and SLS parameters have profound impact on the work contraction. Therefore, this chapter aims at revealing the relationship among the SLS parameters with contraction. It is tiresome to relate parameters in SLS with part contraction by traditional mathematical techniques in a short time. Besides the mathematical models, data models have been employed in AM. These models are known as machine learning (ML).[28,29] The primary benefit of this kind of model is to learn automatically the relationship between the input characteristics and output characteristics from previously available data without any physics-based equations. In various ML techniques, the neural network (NN) is the most popular method. For instance, NNs are applied in areas such as natural language processing[30], voice recognition[31], computer vision[32] and autonomous driving.[33] NN had created a profound effect on all supply chain functions in industry encompassing product design, manufacturing and distribution. The neural network had been applied for function approximation.[34] Three-layered neural network is found to approximate nonlinear continuous and non-singular function.[35] NN exhibits a great ability to recognise the underlying complicated patterns.

Deep learning has exhibited superior abilities to recognise the patterns in the given data applying supervised learning and unsupervised extraction of features.[36] Deep neural network learning is an extended version of the multiple layered perception neural network including a greater number of the hidden layers. It includes better efficient processing of information, more effective computing power and nonlinear activation function. Deep learning has many variants according to various network architectures and training techniques to process various types of data, for example, auto-encoder deep network used for dimensional reduction application.[37] Deep neural network was used to extract probabilistic feature application.[38] Convolutional neural network is characterised by operations like pooling and convolution used for application like pattern recognition and image processing. Recurrent neural networks are used in applications like reasoning of the sequential data. Having data of high amount and large diversity to train the deep neural network, the deep neural network has the capability to pull the features automatically. These can generate a causal relationship about the deep network inputs with the network outputs. Deep neural network learning has become a popular technique applied for natural language processing (NLP),[39] audio processing,[40] computer vision (CV),[41] biomedical technology,[42] autonomous driving[43] and intelligent applications.[44] But the use of deep neural network learning approach is at a very early phase for the additive manufacturing yet.

11.1.1   Aim of the Chapter

This chapter provides a review of the present progress to apply the NN algorithm to AM. The authors consider that a chapter on deep learning application to additive manufacturing will be useful to academicians, researchers and production engineers to provide an understanding about deep learning. Therefore, this chapter aims at helping to learn to apply the deep neural network learning to predict product quality printed by additive manufacturing with an exhibition of the model development and data analytics. In this chapter, deep neural network is shown as the process modelling technique to determine the relationship of the SLS parameters with the part contraction ratio (shrinkage ratio is ratio of the intended value minus the actual value to the intended value). SLS model, developed using deep neural network, has the ability to forecast the amount of contraction resulting in part for a specific set of the SLS parameters. Therefore, prior information about the dimensional accuracy is necessary to take up the actual manufacturing of the product.

The remaining chapter is structured as follows. Section 11.2 provides an introduction to AM technologies. Section 11.3 explains the fundamentals of deep learning techniques. Section 11.4 demonstrates the approach to employ the deep neural network learning to solve the problems in AM. Section 11.5 gives an illustration case. Section 11.6 provides the conclusions.

11.2   Selective Laser Sintering Additive Manufacturing

Powder bed fusion (PBF) is an AM process. It employs a laser thermal energy source in the selective laser sintering additive manufacturing process to melt and fuse selective regions of a powder bed. A typical SLS additive manufacturing process is shown in Figure 11.1. Many preparatory operations are done to make an AM part such that the digital CAD model preparation, STL file generation, STL file import in AM machine and data processing, material powder preparation, the protective atmosphere generation and AM machine warm up. In metal SLS process, the parts are often built on a metallic platform to prevent the thermal strain-developed deformation and for right orientation to reduce need of the support structures. Now, a thin material powder layer is kept on the platform. Next, a cross-section of sliced CAD model is scanned employing a high-power laser beam in order to melt partially and fuse the material powder layer. Here, a laser beam spot travels along with a scanning path. The scanning pattern is produced using the CAD model of the product meant for production. Then, the platform travels down a distance of the layer thickness.

Now, a powder supply platform is moved up in powder supply port to transfer the powder over the build platform. A new material powder layer is

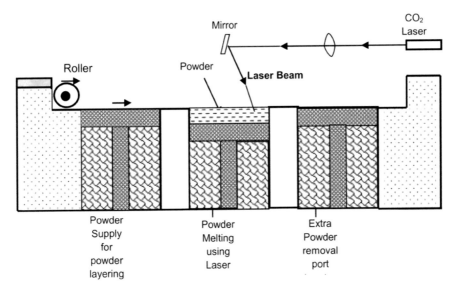

FIGURE 11.1
A typical SLS additive manufacturing process.

formed over the build platform by the roller in the AM machine. A powder removal platform is moved down the powder removal port to remove the excessive metal powder from the machine. These two processes—the powder layering and powder melting—take place alternately till all products are produced. Each layer is joined with the adjacent layers by the laser energy-fused material powder. In end, the parts on the build platform are extracted out of the AM machine. The machine is now cleaned for the next AM part production. Next, the parts on the build platform often undergo thorough heat treatment to remove the thermal stress. Finally, the parts are cut from the platform for performing the post-processing actions like removing the support structure and surface polishing.

Selective laser sintering (SLS) has the merits to recycle the leftover unprocessed metal powder, time efficiency, energy efficiency and geometrical freedom for product design. SLS is a promising AM manufacturing process for various applications such as aerospace and automotive manufacturing.

11.3 Machine Learning

Machine learning is a promising area that employs existing data to predict/respond to the future data. Machine learning is used for computational statistics, pattern recognition and artificial intelligence. Machine learning is vital for fields, for example, spam filtering, facial recognition and other areas with no feasibility/possibility to frame algorithms to do a task. Machine learning is a technique to use machines, that is, computers with software, to find insights from existing data. It also refers to the ability of the machines to learn from the environment. Machines have been employed for human help since the start of civilisation. Machine learning is a process of using an algorithm to transform input data into parameters to interpret the future data. Now, this section describes some key terms used in machine learning.

11.3.1   Data

Every learning technique is based upon data. A data set is employed to impart training to the machine system. Data sets are gathered by people for training. The data set may be of very large size. Machine control systems can gather data using sensors from the system operation. It employs the data to recognise the parameters or to train the machine system.

11.3.2   Models

Models are usually employed in the machine learning systems. The models provide the mathematical framework for the machine learning systems. A model is made by a person. It depends upon the human observations and their experiences. For instance, a model of bus of length 1 and width 1 looking from the top is of a rectangular shape with the size of a standard parking slot. Models are often considered as man-made to give a framework to machine learning. But sometimes machine learning developed its models without any man-made structure.

11.3.3   Training

The machine learning system relates an input data to an output. It is required training to perform this work. Similar to people, as the man needs training to do the tasks, the machine also needs learning systems for their training. Training is provided by feeding the machine system with a known input and the corresponding known output to change the models or data in the learning for relationship learning. Sometimes, it is similar to the curve fitting technique or the regression technique. By having sufficient training data, the machine system gains the ability to generate correct outputs corresponding to new inputs. For instance, on feeding thousands of rat images to the face recognition machine system, the images are generated to be of rats. Now, on giving new rat images, the machine system will be able to identify these as rats. If enough training data sets or the training data in terms of quantity or variety is not provided, it may face a problem in identifying the rats.

11.3.4 Learning Type

This section describes types of learning as below.

Supervised Learning: Supervised learning is a learning in which a particular data set for training is applied to train the machine learning system. Here, the learning is known as supervised, since the training data sets are man-made. It does not show that people validate the results. The process to classify the outputs of the machine system corresponding to given inputs is known as labelling. In other words, one directly indicates whether the results are correct or are there the desired outputs corresponding to every input set. Generating the training data sets is very time-consuming. It requires utmost care to ensure the training data sets to give sufficient training to generate the correct results. These must include the full variety of the inputs and desired outputs. After the training, a set of test data is applied for the validation of results. When the results of the test data aren't of the desired quality, the test data is used as the training set and the process is again repeated. For example, a person is trained exclusively in mechanical engineering. If he was told to work on a civil engineering problem, the results wouldn't be as good as desired since the person did not have the proper training for civil engineering.

Semisupervised Learning: In the semisupervised approach, some portion of data set has the form of labelled training data sets and other data without label. In reality, a small portion of the input data set is labelled data, since labelling requires a skilled person. This small labelled data set is employed to interpret the unlabelled data set.

Unsupervised Learning: Unsupervised learning is a learning type that does not employ the training data sets. It is normally employed to identify the patterns in data set without having any existing right answer. For instance, on employing the unsupervised learning for training a face identification machine system, the machine system may group the data in groups, some groups out of those may be faces. Grouping techniques are often instances of the unsupervised learning. The unsupervised learning enables the machine to learn from the data the things without knowing in advance. It is an important technique to find the hidden structures in data set.

11.4 Deep Neural Network Learning

Deep neural network learning technique evolved from traditional narrow neural network approach. The major difference is that more than two hidden layers are found in the deep neural network over the narrow neural network.

The hidden layers are used to cast the input data on a space of multiple dimensions. Here, the given input data may be examined applying various perspectives . Greater the network hidden layer numbers, greater the probable hidden patterns may be identified in a given data. But increasing layers creates obstacles to train and perform the deep neural network such as the following.

(i) *Severe vanishing gradient problem.* Due increased depth of the neural networks, it becomes difficult for some starting layers of the deep neural network to get the forecasting error. It degrades the effectiveness of the training considerably.

(ii) The risk of the overfitting considerably goes up from increased complexity in the architecture of deep neural network, since it needs more parameters for network training.

(iii) There is increased computation power required to deal with increased complexity of the deep neural network and amount of training data . But these problems had been solved up to some degree with advances in deep neural network learning.

It is made possible by developments such as (i) a new structure of the neural network to have the flow of the loss of information through the entire deep neural network, and to refrain from the gradient disappearance phenomenon, for example, deep neural network ResNet[45]; (ii) robust deep neural network training employing the data sets of large volume alongside a new type activation functions to decrease the problem relating to overfitting; and (iii) use of the faster computation resources like GPU[46]. Therefore, deep learning has proved to be a modern technique for audio processing[26], natural language processing[27] and computer vision[25].

11.5   An Illustration Case

In this section, an example is presented to predict the quality of the printed part to illustrate the application of deep learning for additive manufacturing. As described earlier, it is important to have good quality of the printed part, and therefore, the authors propose to predict it. SLS is a complex additive manufacturing process with many important parameters. Contraction ratio, a measure of part quality, depends upon a number of SLS parameters. The various important SLS parameters [47] are surrounding working temperature (T_e), scanning mode (M_s), layer thickness (l_t), laser power (w), hatch spacing (d_t), scanning speed (v) and interval time (T_s). SLS parameters are to be modelled by a vector X as the following.

$$X = \left[w, l_t, d_t, v, T_s, T_e, M_s \right] \tag{11.1}$$

The contraction ratio Y is to be modelled as function G with argument X:

$$Y = G(X) \tag{11.2}$$

The mathematical model of Equation (11.2) exhibits the effect of SLS parameters on product quality, that is, the shrinkage ratio. When the function G is represented by deep neural network G_n, Y is estimated as

$$Y = G_n(X) \tag{11.3}$$

Here, the deep neural network is capable of expressing relationship of the contraction ratio with the SLS parameter. This chapter has considered a study[47] for the demonstration of deep learning. Specimen material in this chapter is taken as HBI, a composite of polystyrene in the study[47]. The shrinkage ratio[47] is defined as

$$Y\% = \left([S_D - S_M] / S_D \right) \times 100 \tag{11.4}$$

where S_D and S_M are CAD model value and the measured value.

11.5.1   Dataset for the Chapter

The choice of design of the experiment is of paramount for deep learning. The data set for experiments must have a wide representation of the problem [47]. In this chapter, there are the seven process parameters to examine as shown in Table 11.1. The levels of each process parameter were chosen using some references with experience[47].

The SLS process parameters and their levels are displayed for the experiments in Table 11.1. GR-200 analytical balance and AD-1653 density determination kit from A&D Company Ltd. (Tokyo, Japan) [47] is a suitable equipment set to take measurements regarding the density of printed products using SLS. SLS printed part's density was determined corresponding to the product weight in air medium.

11.5.2   Deep Neural Network Parameters

In this chapter, the implementation and training of the deep neural network and its validation were conducted employing TensorFlow library [48] in Python programming language [49]. The deep neural network had seven input nodes for the seven SLS process parameters and the output node has

TABLE 11.1

The SLS Parameters with Levels [47]

Parameters	Levels	Range
Layer thickness (mm)	.1,.16,.2,.24	.1–.24
Laser power (W)	8,8.5,9,10,11,11.5,12,12.5,14,15,18,20	8–20
Scan speed (m/s)	1300,1800,2000,2400,3000	1300–3000
Hatch space (mm)	.08,.1,.12,14,.15	.08–.15
Interval time (s)	0,1,2,3 4	0–4
Surroundings temperature of working (°C)	78,80,84,87,93,95	78–95
Scan mode	1,2	1–2

one node for one response characteristic, that is, SLS manufactured product's relative density. There was 10 hidden layers in the deep neural network. It is found that due to the vast variations in values of seven input SLS parameters and output as product quality characteristic, that is, contraction ratio, the deep neural network learning process resulted in a deep neural network. So, input characteristics as well as the output characteristics should be made in normal form as below.

With mean of characteristic X_i and sample size N, X_{mean} is calculated as,

$$X_{mean} = \frac{\sum_{i=1}^{N} X_i}{N} \tag{11.5}$$

Standard deviation of characteristic X_i is determined as,

$$X_{sd} = \sqrt{\frac{\sum_{i=1}^{N} (X_i - X_{mean})^2}{N}} \tag{11.6}$$

Characteristic X_i is computed in normal form:

$$X_{NormDeviate} = \frac{X_i - X_{mean}}{X_{sd}} \tag{11.7}$$

The rectified linear unit (ReLU), $f(Z_{NormDeviate}) = \max (0, Z_{NormDeviate})$ is an active function, employed for dealing to move from one layer to the next layer [50]. This activation function is a very popular nonlinear function due to its faster learning ability in deep neural networks having multiple layers of neurons [51]. The sigmoid function, an active function, that is, $f(Z_{NormDeviate}) = 1/(1+\exp(-Z_{NormDeviate}))$ was employed for deep neural network's output layer due to output nature of real value. For training of the deep neural network in the operation of hyper parameter optimisation, the average square error is employed as a loss function having Adam optimiser approach [52]. The loss function is computed as

$$E_{od}(w) = \frac{1}{2}\left[\frac{\sum_{i=1}^{N}\left(Y_{Predicted} - Y_{real}\right)^2}{N}\right] \tag{11.8}$$

where Y_{real}, $Y_{predicted}$ and N are the actual magnitude of experimental output, predicted magnitude of the model and grand number of SLS printed parts, respectively.

11.5.3   *K*-fold-Cross-validation for Training the Deep Neural Networks

In order to train the deep neural network model, the data set is needed to be divided into two parts: training data set and test data set. The data set for training had a known output data. The deep neural network model gains learning from this data to predict from other data. The data set for test is employed to check the prediction accuracy of the deep neural network model. But it is very difficult to divide the data set to represent the required problem characteristics. Further, on division of the existing set of data into validation and training data sets, the data set decreases considerably the number of sample of data for use in learning of the model. This is dependent upon a specific random selection of both data sets. So, *k*-fold cross-validation is employed to solve problem specified here. For *k*-fold cross-validation, the given original data set is divided at random in *k*-equal-size data set subsamples. Among *k* subsets of the data sample, one single subsample of data is marked as data set for the validation to validate deep neural network model. Other (*k* – 1) data subsets are employed for training. Now, cross-validation procedure is performed *k* number of times (i.e., folds) considering every *k* subsamples of data set employed one time only as the testing data set for validation. Then, *k* number of results obtained from the folds are taken to compute the average to create a single estimation[53]. The merit of *k*-fold technique is to employ all data sets to train and validate the deep neural network. Every data set is employed to validate one time only. In this chapter, the 10-fold cross-validation approach was employed for training the deep neural network.

11.5.4   Overfitting

Overfitting is a problem that usually happens in machine learning for very accurate model with the training data set; however, it not very accurate with validation data set. On the occurrence of overfitting, the neural network model generally indulges in learning of the noise found within the training data set, rather than actual relationship among the parameters in the data set. In order to refrain from overfitting, the dropout technique and weight decay regularisation technique are employed. Weight decay technique is an important technique to avoid weight to grow very high in value without real requirements. This technique does so with the addition of a term in the loss function to penalise the high weights [54,55] as in the equation below:

$$E_d(w) = E_{od}(w) + \frac{1}{2}.\lambda.w_i^2 \qquad (11.9)$$

where λ is known as a regularisation parameter to control the extent of pen-alisation of weights. W is known as the weight vector representing the entire free variables/parameters related to deep neural network. In the current chapter, the regularisation parameter $\lambda=0.05$ was employed. Drop out is another important technique to avoid the overfitting phenomenon. In the drop out mechanism, units are randomly removed from deep neural network's training phase[56–58]. In the present chapter, the probability of keeping a unit in deep neural network is taken as 0.7.

11.5.5   Results and Discussion

The deep neural network training aims at optimisation of the hyper parameters of deep network model with minimisation of the loss function. For this purpose, the weights are allocated randomly at the beginning. From the given inputs and outputs data for the training, the deep neural network estimates and makes comparison of its predicted outputs with the actual outputs. The computer errors are sent back through the deep neural learning system for adjusting the weights. The process of adjusting the weights is done till the condition of minimal loss is achieved. On completion of the training, the validation test dataset, a part of data set for the experiment, is employed to validate the deep neural network. At last, an average error is computed to obtain grand efficiency of deep neural network. For forecasting performance level evaluation of the deep neural network, an average absolute error (AAE) from Equation (12.10) is used.

$$AAE = \frac{\left|\sum_i^N \left(Y_{predicted} - Y_{real}\right)\right|}{N} \qquad (11.10)$$

Figure 11.2 exhibits the 8-fold-cross-validation strategy for deep neural net-work having 440 epochs for every fold with 0.001 amount of learning rate. Convergence for validation as well as training graphs shows that there is no overfitting by the deep neural network. Further, the value of AAE is 1.53 for the training phase. The value of AAE is 1.54 for the validation phase.

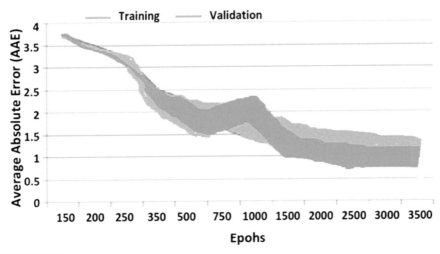

FIGURE 11.2
The 8-fold-cross-validation strategy for deep neural network.

This shows that there is similarity in the average AAE for training as well as validation phase. The value of standard deviation for training phase is 0.21. This lower standard deviation shows that deep neural network does not have any tendency to change considerably with various training subset data. This indicates a good forecasting ability of the deep neural network.

11.6 Conclusions

In this chapter, the quality of a product in terms of minimum shrinkage ratio produced by selective laser sintering was predicted from important SLS parameters such as surrounding working temperature, laser scanning speed, layer thickness, scanning mode, hatch distance, laser power and interval time. The relationship among SLS parameters with contraction ratio is modelled using the deep neural network because the SLS variables are considered to be multitudinous as well as nonlinear. In this chapter, the machine learning system employed supervised deep learning including process parameters as input characteristics and the quality of a product in terms of minimum shrinkage ratio as output characteristic. Weight decay and the dropout technique were employed to overcome the overfitting problem found in deep neural network technique. The predicted outputs characteristic was compared to the actual characteristic. The shrinkage ratio found by the deep neural network can be employed to determine information for shrinkage compensation in the SLS process.

References

[1] Kang, H.S., J.Y. Lee, S. Choi, H. Kim, J.H. Park, and J.Y. Son et al. 2016. Smart manufacturing: past research, present findings, and future directions. *Int J Precis Eng Manuf Green Technol*, 3(January (1)):111–28.

[2] Kusiak, A. 2017. Smart manufacturing must embrace big data. *Nature*, 544 (April (7648)): 23–25.

[3] Lee, J., B. Bagheri, and H.A. Kao. 2015. Cyber-physical systems architecture for industry 4.0 based manufacturing systems. *Manuf Lett*, 3 (January): 18–23.

[4] Lee, J., H. Davari, J. Singh, and V. Pandhare. 2018. Industrial artificial intelligence for industry 4.0-based manufacturing systems. *Manuf Lett*, 18 (October): 20–23.

[5] Kumar, L. and P.K. Jain. 2010. Selection of additive manufacturing technology. *Adv Production Eng Mgmt J*, 5(2): 75–84.

[6] Kumar, L. and P. K. Jain (2006). Rapid prototyping: a review, issues and problems, International Conference on CARs & FOF,VIT, India, 1: 126–138.

[7] Kumar, L. and R. A. Khan (2004). Rapid design and manufacturing. Global Conference on flexible System Management (GLOGIFT) March, JMI, Delhi, 1: 13–15.

[8] Kumar, L., M. Shoeb, and A. Haleem (2020). An overview of additive manufacturing technologies. *Studies in Indian Place Names*, 40(10), 441–450.

[9] Kumar, L., M. Shoeb, A. Haleem, and M. Javaid (2022). Composites in context to Additive Manufacturing, CIMS-2020-International Conference on Industrial and Manufacturing Systems, NIT Jalandhar, Punjab, India, Lecture Notes on Multidisciplinary Industrial Engineering, Springer, Cham, 491–503, ISBN 978-3-030-73494-7.

[10] Kumar L. and P.K. Jain (2022). Carbon conscious and artificial immune system optimization modeling of metal powder additive manufacturing scheduling. *In Computational Intelligence for Manufacturing Process Advancements*, eds. P. Chatterjee, D.P. Željko-Stević, S. Chakraborty, and S. Bhattacharyya, Taylor & Francis, CRC Press (Accepted for publication by Editor).

[11] Lu, B, D. Li, and X. Tian. 2015. Development trends in additive manufacturing and 3D printing. *Engineering*, 1(1):85–89.

[12] Derby, B. 2015. Additive manufacture of ceramics components by ink jet printing. *Engineering*, 1(1):113–123.

[13] Gu, D., C. Ma, M. Xia, D. Dai, and Q. Shi. 2017. A multi scale understanding of the thermodynamic and kinetic mechanisms of laser additive manufacturing. *Engineering*, 3(5): 675–684.

[14] Herzog, D., V. Seyda, E. Wycisk, and C. Emmelmann. 2016. Additive manufacturing of metals. *Acta Mater*, 117(15): 371–392.

[15] Liu, L., Q. Ding, Y. Zhong, J. Zou, J. Wu, and Y.L. Chiu et al. 2018. Dislocation network in additive manufactured steel breaks strength–ductility trade-off. *Mater Today*, 21(4):354–361.

[16] Gorsse, S., C. Hutchinson, M. Gouné, and R. Banerjee. 2017. Additive manufacturing of metals: a brief review of the characteristic microstructures and properties of steels, Ti–6Al–4V and high-entropy alloys. *Sci Technol Adv Mater*, 18(1): 584–610.

[17] Santosa, E.C., M. Shiomia, K. Osakadaa, and T. Laoui. 2006. Rapid manufacturing of metal components by laser forming. *Int J Mach Tools Manuf*, 46(12–13): 1459–1468.

[18] Chen, X., C. Wang, X. Ye, Y. Xiao, and S. Huang. 2001. Direct slicing from power SHAPE models for rapid prototyping. *Int J Adv Manuf Technol*, 17(7): 543–547 doi:10.1007/s001700170156

[19] Li, X.S., M. Han, and Y.S. Shi. 2001. Model of shrinking and curl distortion for SLS prototypes. *Chin J Mech Eng*, 12(8): 887–889.

[20] John, D.W. and R.D. Carl. 1998. Advances in modeling the effects of selected parameters on the SLS process. *Rapid Prototyping J*, 4 (2): 90–96. doi:10.1108/13552549810210257

[21] Yang, H.J., P.J. Huang, and S.H. Lee. 2002. A study on shrinkage compensation of SLS process by using the Taguchi method. *Int J Mach Tools Manuf*, 42(10): 1203–1212. doi:10.1016/S0890-6955 (02)00070-6

[22] Masood, S.H., W. Ratanaway, and P. Iovenitti. 2003. A genetic algorithm for best part orientation system for complex parts in rapid prototyping. *J Mater Process Technol*, 139(3):110–116. doi:10.1016/S0924-0136(03)00190-0

[23] Bai, P.K., J. Cheng, B. Liu, and W.F. Wang. 2006. Numerical simulation of temperature field during selective laser sintering of polymer coated molybdenum powder. *Trans Nonferrous Met Soc China*, 16 (3):603–607. doi:10.1016/S1003-6326(06)60264-1

[24] Arni, R.K. and S.K. Gupta. 1999. Manufacturability analysis for solid freeform fabrication. In: Proceedings of DETC 1999 ASME Design Engineering Technical conference, Vegas, NV, 1–12.

[25] Armillotta, A. and G.F. Biggioggero. 2001. Control of prototyping surface finish through graphical simulation. In: Proceedings of the 7th ADM International conferences, Grand Hotel, Rimini, Italy, 17–24.

[26] Shi, Y., J. Liu, and S. Huang. 2002. The research of the SLS process optimization based on the hybrid of neural network and expert system. In: Proceedings of the International Conference on Manufacturing Automation, 409–418.

[27] Zheng, H.Z., J. Zhang, S.Q. Lu, G.H. Wang, and Z.F. Xu. 2006. Effect of core–shell composite particles on the sintering behavior and properties of nano-Al_2O_3/polystyrene composite prepared by SLS. *Mater Lett*, 60(9–10): 1219–1223. doi:10.1016/j.matlet. 2005.11.003

[28] Kohavi, R. and F. Provost. 1998. Glossary of terms. *Mach Learn*, 30(2–3): 271–274.

[29] Géron, A. 2017. *Hands-on Machine Learning with ScikitLearn and Tensor Flow: Concepts, Tools, and Techniques to Build Intelligent Systems*. Boston, MA: O'Reilly Media Inc.

[30] Devlin, J., M. Chang, K. Lee, and K.B. Toutanova. 2018. Pre-training of deep bidirectional transformers for language understanding. *arXiv*, 1810:04805.

[31] Anusuya, M.A. and S.K. Katti. 2010. Speech recognition by machine- a review. arXiv, 1001–2267.

[32] Krizhevsky, A., I. Sutskever, and G.E. Hinton. 2012. Image Net classification with deep convolutional neural networks. In: F. Pereira, C.J.C. Burges, L. Bottou, and K.Q. Weinberger, eds. Advances in Neural Information Processing Systems 25, Proceedings of Neural Information Processing Systems, 2012, December 3–6, Lake Tahoe, NV, 1097–1105.

[33] Ondruska, P. and I. Posner. 2016. Deep tracking: seeing beyond seeing using recurrent neural networks, *arXiv*, 1602.00991.

[34] Mehrotra, P., J.E. Quaicoe, and R. Venkatesan. 1996. Speed estimation of induction motor using artificial neural networks. *IEEE Trans Neural Netw*, 6: 881–886.

[35] Hornik, K., M. Stinchcombe, and H. White.1989. Multilayer feed forward networks are universal approximators. *Neural Netw*, 2(5): 359–366. doi:10.1016/ 0893-6080(89)90020-8

[36] Lecun, Y., Y. Bengio, and G. Hinton. 2015. Deep learning. *Nature*, 521 (May (7553)): 436–44.

[37] Hinton, G.E. and R.R. Salakhutdinov. 2006. Reducing the dimensionality of data with neural networks. *Science*, 313 (July (5786)): 504–507.

[38] Hinton, G.E., S. Osindero, and Y.W. Teh. 2006. A fast learning algorithm for deep belief nets. *Neural Comput*, 18(7): 1527–1554.

[39] Hirschberg, J. and C.D. Manning. 2015. Advances in natural language processing. *Science*, 349 (July (6245)): 261–266.

[40] Chan, W., N. Jaitly, Q. Le, and O. Vinyals. 2016. Listen, attend and spell: neural network for large vocabulary conversational speech recognition. Proc. IEEE-ICASP, 4960–4964.

[41] Liu, W., Z. Wang, X. Liu, N. Zeng, Y. Liu, and F.E. Alsaadi. 2017. A survey of deep neural network architectures and their applications. *Neuro Computing*, 234(April): 11–26.

[42] Mamoshina, P., A. Vieira, E. Putin, and A. Zhavoronkov. 2016. Applications of deep learning in biomedicine. *Mol Pharm*, 13(May (5)):1445–1454.

[43] Chen, C., A. Seff, A. Kornhauser, and J. Xiao. 2015. DeepDriving: learning affordance for direct perception in autonomous driving. Proc. IEEE-ICCV, 2722–2730.

[44] Covington, P., J. Adams, and E. Sargin. 2016. Deep neural networks for YouTube recommendations. Proc. ACM—RecSys, 191–198.

[45] He, K., X. Zhang, S. Ren, and J. Sun. 2016. Deep residual learning for image recognition. Proc. IEEE-CVPR, 770–778.

[46] Sze, V., Y.H. Chen, T.J. Yang, and J.S. Emer. 2017. Efficient processing of deep neural networks: a tutorial and survey. *Proc. IEEE*, 105 (December (12)): 2295–2329.

[47] Rong-Ji, W., L. Xin-hua, W. Qing-ding, and L. Lingling (2009). Optimizing process parameters for selective laser sintering based on neural network and genetic algorithm. *Int J Adv Manuf Technol*, 42: 1035–1042.

[48] Abadi, M., P. Barham, J. Chen, Z. Chen, A. Davis, and J. Dean et al. 2016. TensorFlow: a system for large-scale machine learning. 12th USENIX Symp. Oper. Syst. Des. Implement. (OSDI' 16). https://doi.org/10.1038/nn.3331

[49] Zelle, J. 2010. Python programming: an introduction to computer science. https:// doi.org/10.2307/2529413

[50] Glorot, X., A. Bordes, and Y. Bengio. 2011. Deep sparse rectifier neural networks. AISTATS' 11 Proc 14th Int Conf Artif Intell Stat, 15: 315–23. doi:10.1.1.208.6449

[51] LeCun, Y.A., Y. Bengio, and G.E. Hinton. 2015. Deep learning. *Nature*, 44: 436–444. https:// doi.org/10.1038/nature14539

[52] Kingma, D.P., and J.L. Adam Ba. 2017. A method for stochastic optimization. Int Conf Learn Represent. https://doi.org/10.1145/1830483.1830503

[53] James, G., D. Witten, T. Hastie, and R. Tibshirani. 2013. *An Introduction to Statistical Learning*, vol. 103. New York: Springer-Verlag. https://doi.org/10.1007/978-1-46147138-7

[54] Zur, R.M., Y. Jiang, L.L. Pesce, and K. Drukker. 2009. Noise injection for training artificial neural networks: a comparison with weight decay and early stopping. *Med Phys*, 36: 4810–4818. https://doi.org/10.1118/1.3213517

[55] Krogh, A., and J.A. Hertz. 1992. A simple weight decay can improve generalization. *Adv Neural Inf Process Syst*, 950–957. doi:https://dl.acm.org/citation.cfm?id=2987033

[56] Srivastava, N., G. Hinton, A. Krizhevsky, I. Sutskever, and R. Salakhutdinov. 2014. Dropout: A simple way to prevent neural networks from overfitting. *J Mach Learn Res*, 15:1929–1958. https://doi.org/10.1214/12-AOS1000

[57] Dahl, G.E., T.N. Sainath, and G.E. Hinton. 2013. Improving deep neural networks for LVCSR using rectified linear units and dropout. ICASSP. IEEE Int. Conf. Acoust. Speech Signal Process. Proc., 8609–8613. https://doi.org/10.1109/ICASSP.2013.6639346

[58] DeRosa, G.H., J.P. Papa, and X.S. Yang. 2018. Handling dropout probability estimation in convolution neural networks using meta-heuristics. *Soft Comput*, 22:6147–6156. https://doi.org/10.1007/s00500-017-2678-4

12

CBPP: An Efficient Algorithm for Privacy-
Preserving Data Publishing of 1:M Micro
Data with Multiple Sensitive Attributes

Jayapradha Jayaram[1,*], Prakash Manickam[2], Apoorva Gupta[3],
and Madhuri Rudrabhatla[4]

[1]*Department of Computer Science and Engineering, College of Engineering and
Technology, Faculty of Engineering and Technology, SRM Institute of Science and
Technology, SRM Nagar, Kattankulathur, Kanchipuram, Chennai, TN, India*

[2]*Department of Computer Science and Engineering, College of Engineering
and Technology, Faculty of Engineering and Technology, SRM Institute of Science
and Technology, SRM Nagar, Kattankulathur, Kanchipuram, Chennai, TN, India*

[3]*Department of Computer Science and Engineering, SRM Institute of Science
and Technology, Kattankulathur, Tamil Nadu, India*

[4]*Department of Computer Science and Engineering, SRM Institute of Science and
Technology, Kattankulathur, Tamil Nadu, India*

*E-mail: jayapraj@srmist.edu.in; prakashm2@srmist.edu.in; as5866@srmist.edu.in;
rr7127@srmist.edu.in*

CONTENTS

DOI: 10.1201/9781003104858-12

12.1   Introduction

Preservation of the individuals' privacy is an essential concern during and after the data is transmitted to third person. In the era of digital world, data owner organizations are discomforted by data privacy. Due to an increase in the generation of data, the task of masking data without disclosing the sensitive information of an individual has become a challenging task. Organizations such as health sectors, pharmaceutical agencies and government sectors often share their data with researchers and third parties for various analysis. Therefore, the data publisher needs to take the responsibility of preserving privacy during data publishing. The data publisher [1] must be a trustworthy person. The data publisher needs to have a complete knowledge about the privacy law and act before disclosing the data to the data recipients. Data publisher should take the responsibility for the privacy of the data and also the utility of the data. Data publisher ensures the data is properly anonymized, so that the sensitive information of the original data is unknown to the data recipient. However, the utility of the original data should be preserved, so that the information needed is acquired correctly and properly. The publisher needs to ensure that the proper model and techniques have been applied on the original data before disclosing the data to the recipient so that there is no clue for sensitive information. Two types of data are of major concern before publishing the data: (i) individual privacy and (ii) collective privacy. Revealing explicit identifiers such as name and id can directly breach a particular individual privacy. To protect personal privacy, the direct identifiers should be removed and the sensitive attributes related to the particular individual need to be made private and anonymized. The safeguarding of an individual privacy may not be adequate. Learning about an individual's sensitive information may also lead to inferring of the information about a group of individuals. Therefore, sensitive knowledge about the data set needs to be preserved. Preserving sensitive knowledge inferred from the data set is termed as collective privacy preservation [2]. The data set consists of two kinds of attributes termed as (i) sensitive attribute and (ii) non-sensitive attributes [3]. A distinctive care needs to be given for sensitive attributes as they contain sensitive information about an individual. These sensitive attributes should not be disclosed to the third party. Non-sensitive attributes in the dataset are projected for the purpose of analysis. The non-sensitive attributes are collectively termed as quasi-identifier. Data publisher discloses the non-sensitive attributes to the third party. However, the non-sensitive attributes can collectively provide the personal information about an individual if it is related with other external sources. Data anonymization should be carried in a proper way by adopting various privacy methods and models. The privacy-preserving models and methods implemented on the data set should balance between the privacy and utility. The basic notion of

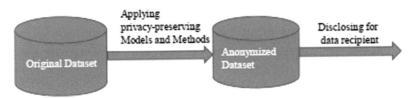

FIGURE 12.1
Privacy-preserved data publishing.

privacy-preserved data publishing is shown in Figure 12.1. The original data set is transformed or anonymized by performing privacy-preserving data publishing models and methods. After anonymization, the data is protected and ready to disclose it to the data recipient [4].

12.2   Related Works

Earlier, the researchers were concentrating on privacy-preserving data publishing on a 1:1 dataset. An individual might have multiple records with different sensitive attributes called a 1:M dataset in the real world. Consider an individual who has cancer. A cancer disease comes with many side effects such as weight loss, vomiting and hair loss. The same individual visits different doctors in the hospital to be treated. So, each time he visits the hospital, a record is registered in the hospital database. Such a scenario leads to a 1:M dataset. When 1:1 privacy models are implemented on 1:M datasets, it leads to various privacy breaches. Several techniques have been proposed for a 1:1 dataset, such as slicing [5], Mondrian [6], suppression [7], clustering and multi-sensitive bucketization [8].

Another method called (k, k_m)-anonymous [9] was proposed, which divided the attributes into relational and transactional attributes. The limitation of (k, k_m) –anonymous is if the information loss for the relational attribute is minimized, the information loss for the transactional attribute would increase and vice-versa. An efficient approach (p,k)-anglicization, was proposed for the anonymization of multiple sensitive attributes. The (p,k)-anglicization provides the optimal balance between privacy and utility [10]. A novel method called overlapped slicing was proposed for privacy-preserving data publishing. Overlapped slicing was implemented with multiple sensitive attributes. Overlapped slicing has proven to provide better utility [11].

The anatomization prevents generalization of quasi-identifiers, thus the information loss is significantly less and results in higher privacy. However, the anatomization method is efficient due to the publishing of multiple tables, and the complexity of the solution increases [12]. Anatomization with slicing is an efficient method used for partitioning the data set with less information loss. The method can preserve privacy, with numerous sensitive attributes [13]. The (QI-MHSA) generalization algorithm has been proposed to anonymize the micro data with multiple sensitive attributes. A vertical partitioning was applied and different models have been adopted to anonymize the categorical and numerical attributes [14]. The L_{sl}-diversity model was proposed for multiple sensitive attributes. Along with L_{sl}-diversity model, three greed algorithms were also proposed to significantly reduce the utility loss of the anonymized published data. Though the utility loss is stable as the data grows, there is a slight increase in their runtime [15].

A model such as decomposition [16,17] was also proposed for privacy-preserving data publishing with multiple sensitive attributes. The slicing and decomposition provide a solution for anonymization. However, they lack utility and have significant information loss. When the slicing is implemented without any anonymization method, it causes considerable information loss; therefore, the suppression of the tuple was combined with sliciVg [18]. (c,k)-Anonymization, an improved method of (p,k)-anglicization was proposed to thwart the fingerprint correlation attack. It improves the one-one correspondence relation in (p,k)-anglicization to one-to-many correspondence [19]. The traditional anonymization techniques such as k-anonymity, l- diversity and t-closeness resulted in optimal results for different conditions [20]. Apart from data anonymization, hiding sensitive attributes has also been implemented in health care, providing better security than traditional and simple methods [21]. The traditional technique k-anonymity cannot be used on high-dimensionality data as it results in a higher information loss [22]. Various techniques have also been implemented and discussed for data anonymization to reduce a high-dimensional dataset into a smaller dimensional data set [23,24]. Later, many researchers focused on 1:M dataset and carried their work to provide an optimal balance between privacy and utility. A new privacy model (k,l)-diversity was proposed to address the re-identification risk in publishing of 1:M dataset. An efficient algorithm called 1:M generalization was proposed to preserve the data utility and privacy. The (k,l)-diversity model was proved to achieve the optimal balance between privacy and utility. The INFORMS and YouTube data set has been used for experimental analysis. However, the privacy model focused on single sensitive attributes [25]. An algorithm named bucket-individual multi-sensitive attribute bucket (QIAB-IMSB) was proposed to achieve an optimal balance between privacy and utility. The ultimate goal of QIAB-IMSB algorithm is anonymize the multi-valued record with less utility loss and high privacy. A vertical portioning was applied on the original data set and implemented k-anonymity on quasi-identifier bucket and (k,l)-diversity on sensitive attribute buckets [26]. A model f-slip was proposed to anonymize the 1:M dataset. An effective anonymization method called "f-slicing" was implemented to

TABLE 12.1

Sample Dataset T^s

| | Explicit Identifier | | Quasi-identifier | | | | | Sensitive Attributes | | |
	Name	Pid	Sex	Age	Zipcode	Disease	Treatment	Symptom	Doctor	Diagnostic Method
tp^1	Avan	1	*	20-30	142**	HIV	ART	Infection	John	Elisa Test
tp^2	Avan	1	*	20-30	142**	Influenza	Medicine	Fever	Alice	RITD Test
tp^3	Avan	1	M	2*	142*	Dyspepsia	Antibiotics	Abdominal Pain	Victor	Ultrasound
tp^4	Becon	2	M	2*	142**	Lung Cancer	Radiation	Weight Loss	Alice	MRI Scan
tp^5	Becon	2	M,F	2*	1420*	Influenza	Medicine	Fever	Alice	RITD Test
tp^6	Canty	3	M, F	28	1420*	HIV	Art	Weight Loss	John	Elisa Test
tp^7	Denny	4	M	25-45	14249	Abdominal Cancer	Chemotherapy	Abdominal Pain	Bob	Chest Xray
tp^8	Emy	5	F	25-45	13084	Covid19	Antibiotics	Fever	Dave	RT-PCR Test
tp^9	Emy	5	M,F	24-45	13084	Asthma	Medication	Chest Tightness	Alice	Methacholine Challenge Test
tp^{10}	Frank	6	M,F	24-45	13064	Asthma	Medication	Shortness Of Breath	Suzan	Methacholine Challenge Test
tp^{11}	Lisa	7	M, F	24-45	13318	Lupus	Medicine	Joint Pain	Jane	Ana Test
tp^{12}	Lisa	7	F	2*	1****	Myocarditis	Medicine	Abnormal Heartbeat	Patrick	ECG
tp^{13}	Ram	8	M	2*	1****	Asthma	Medication	Shortness Of Breath	Suzan	Methacholine Challenge Test
tp^{14}	Ram	8	M	2*	1****	Obesity	Nutrition Control	Eating Disorders	Sana	Body Mass Index

anonymize the sensitive attributes. The f-slip model thwarts five correlation attacks such as (i) background knowledge attack, (ii) multiple sensitive correlation attack, (iii) quasi-identifier correlation attack, (iv) non-membership correlation attack, and (v) membership correlation attack [27]. As per the study, the commonly used 1:M datasets are INFORMS and YouTube [25,26,27]. The chapter is organized as follows: Section 12.3 discusses the contribution of the paper. Section 12.4 validates the various correlation attacks with the scenarios. Section 12.5 discusses the implementation of the CBPP algorithm and Section 12.6 presents the algorithm of CBPP. Section 12.7 describes the evaluation and experimentation results that validate the effectiveness of the CBPP algorithm. Section 12.8 concludes the work with future direction.

12.3   Contributions

The privacy-preserving data publishing models should protect the microdata with high privacy and less information loss. Though various algorithms have been proposed to balance privacy and utility, the challenge the challenge remains unsolved. The significant contributions of the work are as follows:

(1) A detailed study has been accomplished on various privacy-preserving data publishing models of the 1:M dataset.

(2) The anatomization is performed on the original microdata to partition it into two tables.

(3) The quasi-identifier and the sensitive attributes in the microdata are carefully chosen and the distribution of the quasi-identifier is computed.

(4) The generalization process of quasi-identifier and allocation of batch id is done using the CBPP algorithm.

12.4   Correlation Attacks and Those Scenarios

The proposed algorithm can resist three correlation attacks: (i) background correlation attack, (ii) quasi-identifier correlation attack and (iii) non-membership correlation attack. Each correlation attack is discussed with a scenario.

(1) *Attacks using background knowledge correlation*: The intruder can infer the sensitive attributes of an individual if he possesses significant background knowledge about the individual. Case scenario 1 explains the background knowledge correlation attacks.

TABLE 12.2

Anonymity

	Explicit Identifier		Quasi-identifier			Disease	Treatment	Sensitive Attributes		
	Name	Pid	Sex	Age	Zip code			Symptom	Doctor	Diagnostic Method
tp[1]	Avan	1	M	27	14248	HIV	ART	Infection	John	Elisa Test
tp[2]	Avan	1	M	27	14248	Influenza	Medicine	Fever	Alice	RITD Test
tp[3]	Avan	1	M	27	14248	Dyspepsia	Antibiotics	Abdominal Pain	Victor	Ultrasound
tp[4]	Becon	2	M	26	14206	Lung Cancer	Radiation	Weight Loss	Alice	MRI Scan
tp[5]	Becon	2	M	26	14206	Influenza	Medicine	Fever	Alice	RITD Test
tp[6]	Canty	3	F	28	14207	HIV	Art	Weight Loss	John	Elisa Test
tp[7]	Denny	4	M	25	14249	Abdominal Cancer	Chemotherapy	Abdominal Pain	Bob	Chest Xray
tp[8]	Emy	5	F	44	13084	Covid19	Antibiotics	Fever	Dave	RT-PCR Test
tp[9]	Emy	5	F	44	13084	Asthma	Medication	Chest Tightness	Alice	Methacholine Challenge Test
tp[10]	Frank	6	M	45	13064	Asthma	Medication	Shortness Of Breath	Suzan	Methacholine Challenge Test
tp[11]	Lisa	7	F	24	13318	Lupus	Medicine	Joint Pain	Jane	Ana Test
tp[12]	Lisa	7	F	24	13318	Myocarditis	Medicine	Abnormal Heartbeat	Patrick	ECG
tp[13]	Ram	8	M	22	14421	Asthma	Medication	Shortness Of Breath	Suzan	Methacholine Challenge Test
tp[14]	Ram	8	M	22	14421	Obesity	Nutrition Control	Eating Disorders	Sana	Body Mass Index

(2) *Attack using quasi-identifiers correlation*: The intruder can perform a quasi-identifier correlation attack by correlating the quasi-identifier values such as age, zip code, and gender to infer the individual sensitive attributes value and the complete information of the individual. Case scenarios 1 and 2 explain the quasi-identifiers correlation attacks.

(3) *Attacks using non-membership correlation*: The intruder can perform a non-membership correlation attack if he can infer the non-existence of an individual from the data set. The case scenarios 1, 2, and 3 explain the non-membership correlation attacks.

Scenario 1: If the intruder possesses the basic information of an individual, he can gather sensitive information about that particular individual. If the intruder knows that Emy is a female, age>40, from zip code 13084, he can easily infer that Emy falls in either one of the equivalence classes 4 and 5 from Table 12.2. If the intruder also has strong background knowledge that Emy often suffers from breathlessness problems, the probability of inferring Emy's record from equivalence classes 4 and 5 is high.

Scenario 2: If the intruder possesses background knowledge about the individual and the quasi-identifier information, he can easily link the sensitive attributes of an individual by using the quasi-identifier information. If the intruder knows that Becon is a male, age >25, zip code 14206, then the intruder can easily infer that Becon falls in the equivalence classes 2 and 3 from Table 12.2. With the strong background knowledge, he can quickly identify the values of the sensitive attributes of Becon.

Scenario 3: If the intruder has strong background knowledge and possesses the information of quasi-identifier about the individual, that is, Avan is a male, age <30, zip code 14248, also Avan is highly infected with a deadly disease, then he can easily infer that the records of Avan fall in equivalence classes 1 and 2 from Table 12.2. So, the existence of the individual Avan is identified, which leads to privacy breaches. The traditional methods and algorithms cannot thwart the privacy breaches in the 1:M dataset.

12.5 Implementation of the Proposed CBPP Algorithm

Consider a 1:M dataset with multiple sensitive attributes. When the 1:M dataset is generalized using traditional methods such as *k*-anonymity, the intruder can easily infer the sensitive values of the individual, as shown in Table 12.2. The probability of occurrences of privacy breaches in Table 12.2 is also discussed in the previous section.

When traditional models and algorithms have been applied on the 1:M dataset, it may cause privacy breaches. Therefore, an efficient CBPP algorithm has been proposed to resist background knowledge attacks, non-membership correlation attacks and quasi-identifier correlation attacks. CBPP algorithm partitions the original data set into two tables: the quasi-identifier table and the sensitive attribute table. Both are linked together by a batch id, which is further explained below.

Consider the original sample dataset T^s in Table 12.1. When the algorithm CBPP is applied on the data set, it converts the original data set into two tables. The original dataset T^s consists of multiple sensitive attributes for a particular individual and shares the same quasi-identifiers such as age, zip code, and gender. Patient ID is just used for reference, and it will be removed during the publishing of data. The records having same patient id are merged. For example, the first three records of Table 12.1 with patient id 1 are merged. (i.e., $tp^1 U\ tp^2 U\ tp^3$). After merging the records of an individual, an aggregated table T^{sa} is formed. Anatomization is performed on the aggregated dataset T^{sa} and partitions the data set into two tables: (1) quasi-identifier T^{sq} and (2) sensitive attribute T^{ss}. Let q^d be the quasi-identifier values of table T^{sq} and the set of quasi-identifier attributes are $(q^d_1, q^d_2, q^d_3 \dots q^d_n)$. The quasi-identifier table T^{sq} comprises age, sex, and zip code.

Let s^d be the values of the sensitive attributes of T^{sa} and the set of sensitive attributes are $(s^d_1, s^d_2, s^d_3 \dots s^d_n)$. The sensitive attributes are disease, treatment, symptom, doctor, and diagnostic method. The quasi-identifier attribute values are generalized and a batch id is allotted, as shown in Table 12.3. The sensitive attribute table is formed as shown in Table 12.4 according to the batch id. During the anonymization process, the data gets shuffled. Thus, batch id, written as b^{id}, is allocated to link the records in both quasi-identifier and sensitive attribute table.

TABLE 12.3

Quasi Identifier Table T^{sq}

Sex	Age	Zipcode	Batch Id
M	25-45	13000-15000	2
M	25-45	13000-15000	2
F	25-45	13000-15000	2
M	25-45	13000-15000	2
F	41-60	13000-15000	3
M	41-60	13000-15000	3
F	15-24	13000-15000	1
M	15-24	13000-15000	1

TABLE 12.4

Sensitive Attribute Batch Table T

Disease	Treatment	Symptom	Physician	Diagnosis Method	Batch id
Influenza, HIV, asthma, obesity, lupus, myocarditis	Art, medicine, medication, nutritional control	Infection, fever, shortness of breath, eating disorder, joint pain, abnormal heartbeat	John, Alice, Suzan, Sana, Jane, Patrick	ELISA Test, RITD Test, methacholine challenge Test, body mass index, Ana Test, ECG	1
HIV, influenza, dyspepsia, lung cancer, abdominal cancer, asthma	Art, medicine, radiation, chemotherapy, medication	Infection, fever, abdominal pain, weight loss, chest tightness, shortness of breath	John, Alice, Victor, Bob, Suzan,	ELISA Test, RITD Test, Ultrasound, MRI scan, chest X-ray, methacholine challenge test	2
Phthisis, asthma, influenza, HIV, dyspepsia, abdominal cancer	Antibiotics, medication, medicine, radiation, art, chemotherapy, medication	Fever, chest tightness, shortness of breath, infection, abdominal pain	David, Suzan, Alice, John, Victor, Bob, Dave	Molecular diagnostic methods, methacholine challenge test, RITD test, ELISA test, ultrasound, chest X-ray, RT-PCR test	3

12.5.1 Generalization and Allocation of Batch Id

The generalization process replaces the original value into a particular range of values (e.g.) if an individual's age is 25, then it can be replaced in the interval of 20–30. The allocation of batch id prevents the privacy breaches as the tables to be published only with batch id. The batch id, b^{id}, is allotted by using the combinations of quasi-identifiers.

Assume that we generalize the age as 25–30, 30–35, and so on. So, the batch id is allotted to each group as 1, 2 up to n if there are n distributions in a particular attribute. In the sample data set, three different quasi-identifiers are present. To find the total number of possible combinations, let's say that values of sex age and zip code are distributed in three groups. Therefore, the total number of batch id combinations possible is 2*3*1+3 = 9. Out of these 9 possible combinations, any combination can be used. When the number of combinations is high, it can be decided by the user. Therefore, it becomes challenging to find the individual in the possible combination. Hence, batch id helps in preserving the privacy of the data. The complete architecture of the CBPP algorithm is shown in Figure 12.2.

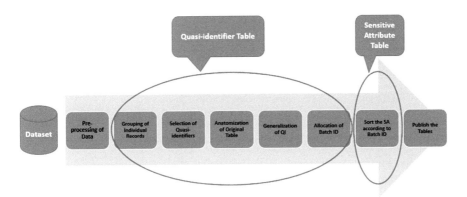

FIGURE 12.2
Architectural diagram of the CBPP algorithm.

12.6 CBPP Algorithm

The algorithm has been divided into two parts and each part is elaborated. In algorithm 1, the original data set is passed as an input argument in line 1. In line 2, the multi-records of an individual are merged using the group by function such that each individual has only one entry. In line 3, anatomization has been performed to partition the original table T^s into two tables: (a) quasi-identifiers (T^{sq}) and (b) sensitive attribute table (T^{sa}). In line 4, the quasi-identifier is identified as sex, age, and zip code. The new list for age and zip code are created in lines 5 and 6. In line 7, a new list has been created for batch id. If the value for age ranges from 15 to 60, the optimal number of distributions is 3, that is, 15–24, 25–44, and 45–60. From lines 8 to 10, the quasi-identifier age has been generalized. For example, if the person's age is 23, then the value 15–24 will be appended to the list. Similarly, in lines 11–13, the quasi-identifier zip code, distributions are formed and values are appended into the zip code list stated in lines 14 and 15. In the sample data set of the work, the zip code ranges from 13000 to 15000. So, in Table 12.3, the distribution of zip code ranges from 13000 to 15000 for all the individuals. According to the algorithm, the total number of batch id combinations possible is 2*3*3+3 = 21, whereas the original sample data set of the work has the total number of batch id combinations possible as 2*3*1+3 = 9.

In algorithm 2, sensitive attributes are arranged according to the batch id and the quasi-identifier that has been used. In line 1, the anatomized sensitive attribute table has been passed as an argument. From lines 2 to 4, three lists of each sensitive attribute have been created. Specifically, three lists are created because the number of distributions that have been made in the quasi-identifiers is 3 and that quasi-identifier in return gives existence to batch id. In line 5, iterate the values according to age and each value is sent to the list corresponding to the batch id. A similar procedure is performed for all the sensitive attributes and the sensitive values are clubbed together according to batch id.

Algorithm 1: Creating quasi-identifier table from original aggregated table
1. Input(T^s)
2. T^{sa} ← T^s.groupby(patient_id)
3. anatomization(T^{sa})
4. T^{sq} ← T^s ["sex", "age", "zipcode"]
5. age_class ← []
6. zip_class ← []
7. Batch_id ← []
8. For x in range (length(T^{sq})
   a ← T^{sq}.loc(i, age)
   If(a > 15) && (a < 25)
   age_class.append(15-24)
   b^{id}.append(1)
9. If(a>=25) && (a<=40)
   age_class.append(25-40)
   b^{id}.append(2)
10. If(a>=41) && (a<60)
    age_class.append(41-60)
    b^{id}.append(3)
11. for y in range(length(T^{sq})):
    b ← df_gen.loc[i, 'ZIPCODE']
    if (b < 10000) and (b > 13000):
    zipcode_class.append("10000-13000")
12. elif (b >=13000) and (b < 16000):
    zipcode_class.append("13000-16000")
13. else:
    zipcode_class.append("16000-19000")
14. T^{sq}.drop(age, zipcode)
15. T^{sq} (age_class,zip_class)

Algorithm 2: Creating sensitive attribute from aggregated table
1) Input(T^{sa})
2) s^{d1}_list ← []
3) s^{d2}_list ← []
4) s^{d3}_list ← []
5) for i in range(length(df)):
   age ← df.loc[i, age]
   if (age >= 15) and (age < 26):
   # For Disease
   d ← df.loc[i, 'disease']
   slist ← s.split(',')
   for dis in slist:
   if dis not in dis_1:
   dis1.append(dis)
6) END

12.7   Evaluation

The CBPP algorithm is implemented in Python and the experiments are conducted in the machine that runs on Windows 10, 8 GB RAM, 1 TB storage and 128 GB SSD. The CBPP algorithm is performed on the real-world 1:M dataset INFORMS. There are 2, 30,231 records in INFORMS dataset. After grouping of records and removal of duplicates, the size of the dataset is 40,126. The birth year, month, sex, and race are chosen as quasi-identifiers and education year, income, and poverty line are taken as sensitive attributes. The information loss is measured using query accuracy.

12.8   Measurement of Information Loss

12.8.1   Query Accuracy

Since the information loss needs to be measured from both the tables, (i) quasi-identifier and (ii) sensitive attribute, usual methods of utility measurement cannot be performed. Hence, we use query accuracy for the calculation utility of the algorithm. The query accuracy measures the information loss with the result of aggregate queries.

$$\text{Query Error} = \frac{\left|\sum(QI) - \sum(\text{Org})\right|}{\text{count}(\text{Org})} \qquad (12.1)$$

The query error measures the COUNT queries executed on the micro-table data set and original data set for measuring information loss. In this case, the total number of batch ID combinations possible is 9. Out of these 9, only one particular combination is the key that links the quasi-identifier table and the sensitive batch table. Hence, batch id is the first line of protection. The probability that the attacker successfully finds the perfect combination becomes 1/9 = 0.11, which is very low. The combinations of batch id can be increased by making more distributions in age and sex, further decreasing the probability of identifying the batch id combination. In Figure 12.3, it is clearly shown that information loss is inversely proportional to the number of values. Therefore, information loss would decrease with the increase in the number of records. However, in k-anonymity, information loss increases as the dimensionality increases [18], as shown in Figure 12.3. The CBPP algorithm provides better utility when compared with the traditional method, k-anonymity.

Consider that, in some cases, the attacker can get through the batch ID. Then the intruder further has to make proper combinations of all the shuffled

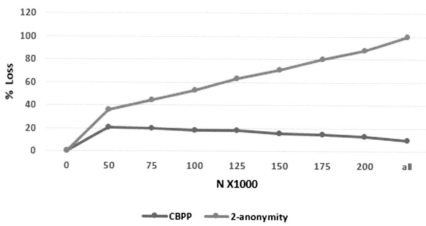

FIGURE 12.3
Information loss.

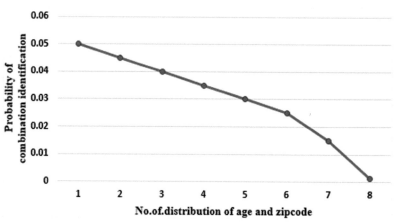

FIGURE 12.4
Probability of identification of quasi-identifier combination versus number of distributions.

sensitive values to obtain the proper and complete details about the individuals. The relation between the distributions and the probability to crack the combination is given in Figure 12.4.

12.9   Conclusion and Future Directions

The study presents an efficient algorithm for privacy-preserving data publishing of the 1:M data set. The proposed CBPP algorithm addresses the solution for three correlation attacks such as background correlation attack, quasi-identifiers correlation attack, and non-membership correlation attack. The case scenarios in the chapter explain that the traditional method cannot be applied to the 1:M dataset. The probability of privacy breaches was determined and the information loss is measured and compared with *k*-anonymity. Experiments were carried out on the real-world dataset INFORMS. The query accuracy is used as a measure for information loss to show that CBPP is providing less information loss compared to the traditional method. Still, there are contents to be worked on for future directions. The probability of identifying the combinations upon the distribution of quasi-identifier using batch id should be implemented on a higher dimension data set. Also, the study focuses on static QID, whereas the quasi-identifier can also change as time changes.

References

[1]   Rashid A. and Yasin N., Privacy preserving data publishing: review. *International Journal of Physical Sciences*, 10(7), 239 (2015).

[2]   Sheppard, E., Data Privacy is a collective concern, 2020. https://medium.com/opendatacharter/data-privacy-is-a-collective-concern-8ebad29b25ce

[3]   Vasudevan L., Sukanya S., and Aarthi N., Privacy preserving data mining using cryptographic role based access control approach, Proceedings of the International MultiConference of Engineers and Computer Scientists, Hong Kong, March 19–21, 2008, p. 474.

[4]   Churi Prathamesh, P. and Pawar Ambika, V., A systematic review on privacy preserving data publishing techniques, *Journal of Engineering Science and Technology Review*, 12(6), 17–25 (2019).

[5]   Li, T., Li, N., Zhang, J., and Molloy, I. Slicing: a new approach for privacy-preserving data publishing, *IEEE Transactions on Knowledge and Data Engineering*, 24(3), 561–574 (2010).

[6]   Wantong Zheng, Zhongye Wang, Tongtong Lv, Yong Ma, and Chunfu Jia, K-anonymity algorithm based on improved clustering, *International Conference on Algorithms and Architectures for Parallel Processing*, 11335, 426–476 (2018).

[7]   Elanshekhar, N. and Shedge, R. An effective anonymization technique of big data using suppression slicing method, International Conference on Energy, Communication, Data Analytics and Soft Computing (ICECDS), 2500–2504 (2017).

[8]   Radha, D. and Valli Kumari Vatsavayi, Bucketize. Protecting privacy on multiple numerical sensitive attribute, *Advances in Computational Sciences and Technology*, 10(5), 991–1008 (2017).

[9]   Puri, V., Sachdeva, S., and Parmeet Kaur, Privacy preserving publication of relational and transaction data: Survey on the anonymization of patient data, *Computer Science Review*, 32, 45–61 (2019).

[10] Anjum, A., Ahmad, N., Malik, U.R., Zubair, S., and Shahzad, B., An efficient approach for publishing micro data for multiple sensitive attributes, *The Journal of Supercomputing*, 74, 5127–5155 (2018).

[11] Widodo, Budiardjo, E.K., and Wibowo, W.C. Privacy preserving data publishing with multiple sensitive attributes based on overlapped slicing. *Information*, 10, 362 (2019).

[12] Lin Yao, Zhenyu Chen, and Xin Wang, Dong Liu, and Guowei Wu, Sensitive label privacy preservation with anatomization for data publishing, *IEEE Transactions on Dependable and Secure Computing*, 18(2), 904–917 (2019).

[13] Susan, V.S. and Christopher, T. Anatomisation with slicing: a new privacy preservation approach for multiple sensitive attributes. *SpringerPlus* 5, 964 (2016).

[14] Jayapradha, J., Prakash, M., and Harshavardhan Reddy, Y., Privacy preserving data publishing for heterogeneous multiple sensitive attribute with personalized privacy and enhanced utility, *Systematic Reviews of Pharmacy*, 11(9), 1055–1066 (2020).

[15] Yuelei Xiao and Haiqi Li, Privacy preserving data publishing for multiple sensitive attributes based on security level, Information, MDPI, 11, 1–27 (2020).

[16] Das, D. and Bhattacharyya, D.K., Decomposition: improving *l*-diversity for multiple sensitive attributes, International Conference on Computer Science and Information Technology 403–412 (2012).

[17] Yang Ye, Liu Yu, Chi Wang, Depang Lv, and Jianhua Feng, Decomposition: privacy preservation for multiple sensitive attributes. International Conference on Database Systems for Advanced Applications, 486–490 (2009).

[18] Kiruthika, S. and Mohamed Raseen, M., Enhanced slicing models for preserving privacy in data publication, International Conference on Current Trends in Engineering and Technology, IEEE, 1–8 (2013).

[19] Khan, R., Tao, X., Anjum, A., Sajjad, H., Khan, A., and Amiri, F. Privacy preserving for multiple sensitive attributes against fingerprint correlation attack satisfying c-diversity. *Wireless Communications and Mobile Computing*, 1–18 (2020).

[20] Bennati, S. and Kovacevic, A., Privacy metric for trajectory data based on *k*-anonymity, *l*-diversity and t-closeness, (2020), https://arxiv.org/pdf/2011.0921 8v1.pdf

[21] Pika, A., Twynn, M., Budiono, S., Ter Hofstede, A.H.M, van der Aalst, wil MP, and Reijers, H.A. Privacy-preserving process mining in healthcare, *International Journal of Environmental Research Public Healthcare*, 17(5), 1612 (2019).

[22] Aggarwal, C.C. On *k* anonymity and curse of dimensionality, 31st International Conference on Very Large Databases, 433–460 (2018).

[23] Bild, R., Kuhn, K.A., and Prasser, F., Better safe than sorry—implementing reliable health data anonymization, *Student Health Technology Inform*, 270, 68–72 (2020).

[24] Wang, R., Zhu, Y., Chen, T., and Chang, C. Privacy-preserving algorithms for multiple sensitive attributes satisfying t-closeness. *Journal of Computer Science Technology* 33, 1231–1242 (2018).

[25] Qiyuan Gonga, Junzhou Luo, Ming Yang, Weiwei Ni, and Xiao-Bai Li, Anonymizing 1:M microdata with high utility, *Knowledge-Based Systems*, 1–12 (2016).

[26] Jayapradha, J. and Prakash, M. An efficient privacy-preserving data publishing in health care records with multiple sensitive attributes, 6th International Conference on Inventive Computation Technologies, 623–629 (2021).

[27] Jayapradha, J. and Prakash, M, f-slip: An efficient privacy-preserving data, publishing framework for 1:M microdata with multiple sensitive attributes, Soft Computing, pp. 1–18, (2021).

13

Classification of Network Traffic on ISP Link and Analysis of Network Bandwidth during COVID-19

V. Ajantha Devi,[1] **Yogendra Malgundkar,**[2] **and Bandana Mahapatra**[2]

[1]*Research Head, AP3 Solutions, Chennai, India*

[2]*Symbiosis Skills and Professional University, India*

E-mail: ap3solutionsresearch@gmail.com

CONTENTS

DOI: 10.1201/9781003104858-13

13.1 Introduction

Nowadays, ISP networks are more complex than before, the number of applications that run on clients/servers are increasing leading to a network resource management problem. The IT professionals tasked with maintaining the network are facing serious challenges in determining which applications consume the resources or degrade network performance. Analyzing the performance metrics collected from clients/servers using traditional methods is not enough anymore to make a correct decision whether an application consumes more network resources. The improvement of the network performance is not an easy task, and even a network bandwidth upgrade might not be an optimal solution to solve the problem of high network utilization.

In this study, taking advantage of machine learning, business intelligence and data analytics to compare and show the benefits of using those techniques in analyzing the network performance metrics, performance metrics from a real ISP network are collected. A cleaned data with performance metrics will be analyzed to find the pattern and correlation to give a better understanding of the applications and network performance. The data set will be used to apply machine learning algorithms to predict future network performance under certain conditions. The implementation of those techniques in future will cut the cost of running a network and reduce investigation time whenever a problem occurs and make IT professional's life much easier.

Due to the COVID-19 situation, almost 100% workforce across organizations and various businesses were working from home/remotely. This results in heavy utilization of internet bandwidth at the ISP side. As a result, it's critical for ISPs to understand and classify the many types of network traffic that pass through their network. Internet service providers can use this classification to successfully control the network performance of their Internet lines and deliver a high-quality service to their consumers, leading to high customer satisfaction.

Network traffic classification is also significant for network security and for determining intrusion detection, QoS, and other features. Forecasting internet bandwidth demand would also greatly assist ISPs in efficiently planning their network resources. ISPs are interested in centralized measures and detecting problems with specific customers before the they complain about the difficulties, and if possible, before the consumers discover the problems at all.

13.2 Methodology

An experiment will be conducted out to ascertain the effects of applications on resource consumption in the networks by extracting performance metrics. The data is analyzed using machine learning and data analytics techniques, in addition to finding any correlation that exists between those metrics. The performance metrics are extracted from the client as well as the server. The clients are accessing files or web applications over the internet links.

The methodology is outlined as follows:

- Data collection
- Parsing the data—data cleaning, scaling, and normalization as required
- Generating graphs—to find trends and patterns in utilizations, find correlations
- Data modelling using ML techniques to predict the class of network traffic
- Predictive modelling to forecast the bandwidth requirement
- Analysis technique: would include Machine Learning classifiers.

Data Collection Method: This is the most crucial and first step, which comprises data collection. The real-time network traffic is recorded in this step. There are several programs for capturing network traffic; however, the TCP dump utility can capture real-time network information. The solar wind packet capturing and analysis program is utilized to gather network traffic. Application traffic such as WWW, DNS, FTP, P2P, and Telnet is recorded. Random network connection details were captured for the duration of one year as shown in Figure 13.1.

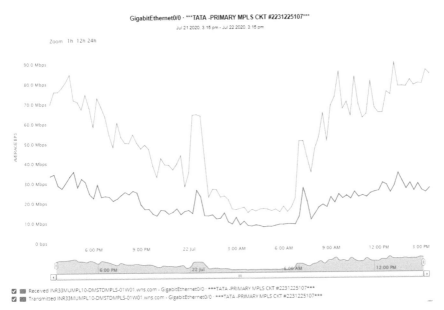

FIGURE 13.1
Random network connection captured in mid-July 2019 to mid-July 2020.

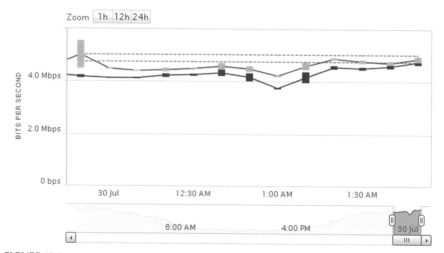

FIGURE 13.2
Network traffic data is collected for one of the links from the ISP.

Network traffic data is collected for the defined duration for one of the links from the ISP as shown in Figure 13.2. This is done by masking client-specific information to avoid the compliance issues.

Data Set Information: The following attributes are available in a data set for analysis.

- Network traffic classification data set

Variable	Description
Flow ID	Information about end-to-end network traffic generated
Source IP	IP address of the source host
Source Port	Network port number of sources
Destination IP	IP address of the target host/server
Destination Port	Network port number of destinations
Protocol	Network protocol number
Protocol Type	Type of protocol: TCP / UDP
L7 protocol	OSI Layer seven protocol used in the connection
Application name	Target application, which is accessed in network session

A total of 1048575 observations were made with nine network attributes.

- Network bandwidth data set

Variable	Description
Timestamp	Date and time of the details captured
Average TX Mbps	Average transmit speed on the link on that day in Mbps
Average RX Mbps	Average receive speed on the link on that day in Mbps

Variable	Description
Peak TX Mbps	Peak transmit speed on the link on that day in Mbps
Peak RX Mbps	Peak receive speed on the link on that day in Mbps
Max_users	Maximum number of user connections on that day
Bandwidth	Bandwidth utilized on the link in a day

A total of 356 observations with 7 attributes including time series details. Bandwidth details were captured for one year—from July 17, 2019, to July 6, 2020, on a daily basis.

13.2.1 Network Topology

The network and application performance is measured through several performance metrics such as bandwidth, throughput, disk time, number of packets send/Recv per sec, and number of bytes send/Recv per second. The network used in the experiment has several tools to collect network and application metrics as in Figure 13.3. The metrics are collected on the hosts and the communication link between the two end points.

The tables below show the performance metrics categorized by the tool used:

Network monitor	Performance monitor
Flow ID	Bandwidth
Source IP	Transfer rate
Destination IP	Receive rate
Source port	

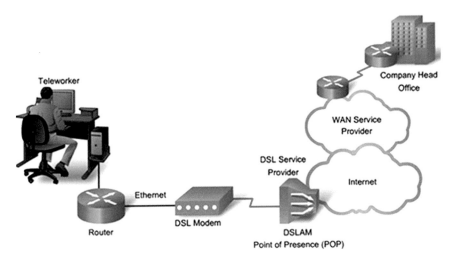

FIGURE 13.3
Network performance metric tools.

Network monitor	Performance monitor
Destination port	
Protocol	
Protocol type	
L7_protocol	
Application name	
Max_Users	

13.2.2   Data Cleaning and Preprocessing

Preprocessing is a crucial step in transforming real-world data into a usable format. Certainly, real-world data is frequently incomplete and noisy in specific behavior. To put it another way, most of the data needed to analyze using data mining techniques comes from the real world and is incomplete and inconsistent (containing errors, outlier values). As a result, pre-processing procedures are required before using data mining techniques to improve the data quality, thereby contributing to improving the accuracy and efficiency of the resultant data mining task. Due to the patterns of network traffic, which have diverse types of format and dimensions, pre-processing techniques are necessary and important in network traffic analysis. All the categorical variables are converted to numeric variables. All the categories are converted into numeric levels except "Application Name," which is the target variable for classification.

13.2.3   Data Visualization with Tableau

It's observed that Google, HTTP, and SSL connections are the top five applications in the network traffic as shown in Figure 13.4. YouTube, Windows updates, and Skype also are high contributors to the traffic flow on the internet link.

From the above trend, it is shown that bandwidth requirements increased almost exponentially from March 2020 onwards as shown in Figure 13.5. This can be related with lockdown situations due to COVID-19 where most of the workforce of many organizations and institutions were forced to work from home. This has increased the internet utilization across the globe. The number of internet users increased drastically from 1 April 2020 onwards as shown in Figure 13.6. As much as 7,000 to 12,000 users were connected in the period of May to July 2020.

Peak transmit and receive speed is in line with the bandwidth usage as observed for the one-year period as shown in Figure 13.7. Transmit speed is seen high increase during lockdown period.

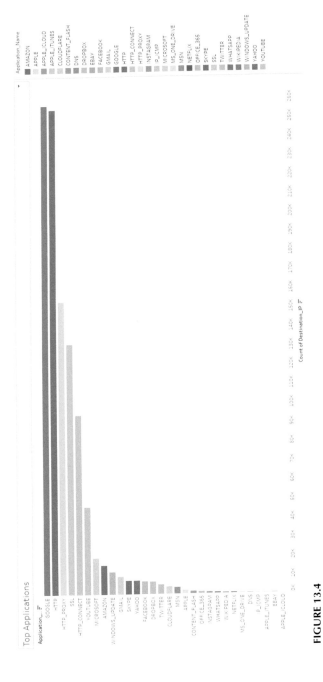

FIGURE 13.4
Top five applications in the network traffic.

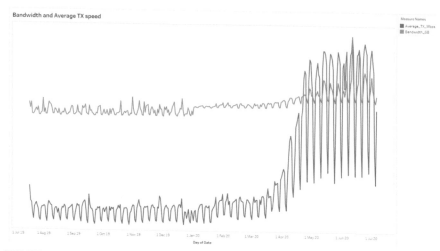

FIGURE 13.5

Traffic increased almost exponentially from March 2020.

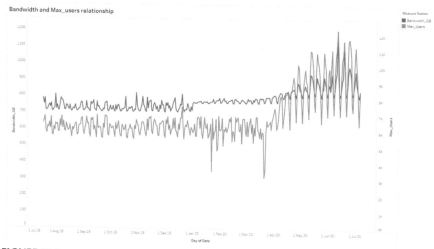

FIGURE 13.6

Increased internet users from April 1, 2020.

13.3　Traffic Classification Using Classification Algorithms

The process followed for network traffic classification model is as follows (Figure 13.8):

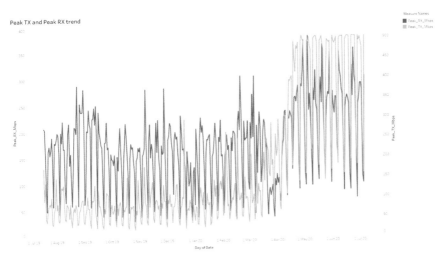

FIGURE 13.7
High increase during lockdown period.

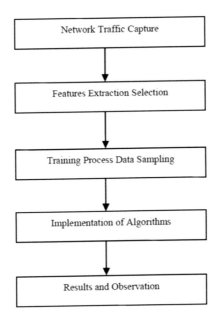

FIGURE 13.8
Network traffic classification model.

The cleaned data is set ready for classification as follows.

```
1  Network.head()
```

	Flow.ID	Source_IP	Source_Port	Destination_IP	Destination_Port	Protocol	Protocol_Type	L7_Protocol	Application_Name
0	33547	13	80	34	36240	6	1	7	18
1	432768	25	38687	5757	80	6	1	126	16
2	8311	25	57735	6684	80	6	1	7	18
3	147160	25	35318	3028	443	6	1	126	16
4	10621	25	33607	8733	443	6	1	178	0

13.3.1 Comparison of Classifiers

The simulation tool provides precise results regarding the applied methods, such as accuracy, training time, and recall, after various machine learning algorithms have been implemented. The four different classifiers used in this exercise are KNN, decision tree, random forest, and naïve Bayes.

When compared to other algorithms, the random forest algorithm produces extremely accurate results. The accuracy results of applying these machine learning techniques are compared in the chart in Figure 13.9.

Classification accuracy of 69 percent has been achieved using Naïve Bayes algorithm. With $k=4$, the classification accuracy of 89 percent is achieved with KNN algorithm. Setting $k=3$ and running the code again gives us better results as shown below. Accuracy increased to 91 percent when the value of k is reduced from 4 to 3. The decision tree gives improved accuracy of 92.89 percent compared to KNN. Random forest gives the highest accuracy of 99.9 percent among all the algorithms attempted in classifying the network traffic.

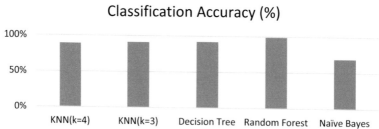

FIGURE 13.9
Applying the accuracy results.

13.4   Bandwidth Requirement Prediction Using Time Series Model (ARIMA)

The time series data of bandwidth utilization for one of the ISP links is given in the following table:

TimeStamp	Average TX (Mbps)	Average PX (Mbps)	Peak TX (-Mbps)	Peak RX (Mbps)	Max Users	Bandwidth (GB)
2019-07-17 15:00:00	71.75	117.1	172.6	207.39	6509	782.06
2019-07-18 15:00:00	47.06	106.67	90.46	204.74	6661	747.22
2019-07-19 15:00:00	44.98	36.31	135.53	153.86	6895	781.23
2019-07-20 15:00:00	27.96	44.11	45.54	94.64	6052	711.90
2019-07-21 15:00:00	17.48	30.51	36.36	48.91	5864	707.87
2019-07-22 15:00:00	36.77	81.78	64.52	163.03	6309	716.80
2019-07-23 15:00:00	41.96	97.95	77.84	173.13	6217	733.06

There are no null values as this is a TCP dump from the tool. All the network parameters captured for the specified time stamp are given below.

```
1   Bandwidth.isnull().sum()

timeStamp          0
Average_TX_Mbps    0
Average_RX_Mbps    0
Peak_TX_Mbps       0
Peak_RX_Mbps       0
Max_Users          0
Bandwidth_GB       0
dtype: int64
```

13.4.1   Feature Extraction Selection

Before using data mining techniques, feature extraction selection is used as a pre-processing strategy. It is used to increase the performance of data mining algorithms by removing redundant or irrelevant features. By selecting only a subset of the original characteristics, feature extraction selection methods generate a new set of attributes. Its primary purpose is to minimize the dimensionality of a data set in order to improve network traffic analysis. The correlation matrix of the variables is obtained in a data set to identify the relevant features for bandwidth prediction. From the correlation coefficients, the following can be observed:

- High positive correlation between average TX speed and bandwidth
- High positive correlation between peak TX speed and bandwidth
- High positive correlation between maximum users and bandwidth
- Moderate relationship between RX speeds and bandwidth.

Hence transmit speed parameters and user count are used as features in the prediction model.

- Average TX speed (Mbps)
- Peak TX speed (Mbps)
- Maximum users

The RX parameters was dropped from the data set for further modelling. Then timestamp column was set as an index for the data set, which is required for time series modelling.

```
1  df['timeStamp'] = pd.to_datetime(df['timeStamp'])
```

```
1  df.head()
```

	timeStamp	Average_TX_Mbps	Peak_TX_Mbps	Max_Users	Bandwidth_GB
0	2019-07-17 15:00:00	71.75	172.66	6509	782.06
1	2019-07-18 15:00:00	47.06	90.46	6661	747.22
2	2019-07-19 15:00:00	44.98	135.63	6895	781.23
3	2019-07-20 15:00:00	27.96	45.54	6052	711.90
4	2019-07-21 15:00:00	17.48	36.36	5864	707.87

```
1  BW_df=df.set_index('timeStamp')
```

The subplots of all the variables with respect to index (timestamp) provides insights into data variation and trend as shown below in Figure 13.10.

There is substantial increase in network throughput post April 2020. This is exactly when most of work started happening online across the industries. This has put a heavy load on ISPs as almost all entire workforce across the world is working remotely as offices were shut.

13.4.2   Auto-Regressive Integrated Moving Average Model (ARIMA)

In order to predict the network traffic, the ARIMA model is used in conjunction with a time series model. To anticipate the bandwidth demand, we'll

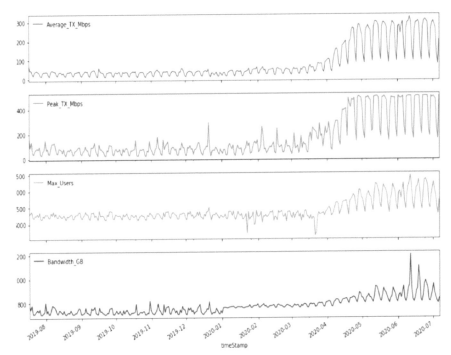

FIGURE 13.10
Subplots of index of timestamp.

utilize the ARIMA module in Python to create a time series model. It's critical to examine the trends and seasonability of the time series data before moving further with the TS model construction.

In comparison to 2019 and 2020, there is a significant difference in bandwidth utilization as shown in Figure 13.11.

13.4.3   Time Series Data: Trends and Seasonality

Finding trends in a time series can be done in a variety of ways. A rolling average is a common technique that includes taking the average of the points on either side of each time point. It's worth mentioning that the number of points is influenced by the size of the window chosen. Noise and seasonality are smoothed out as a result of utilizing the average. Considering a window of 15 days, the trend line of bandwidth usage also shows a gradual rise with time. As a result, when compared to the prior plots in Figure 13.11, the majority of the seasonality has been erased as in Figure 13.12. The same plots for TX speeds can be visualized in Figure 13.13 and number of users connected to ISP link in Figure 13.14.

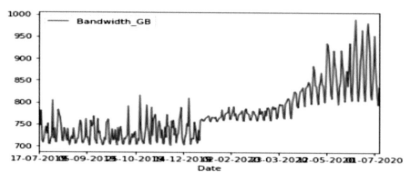

FIGURE 13.11
Difference in bandwidth.

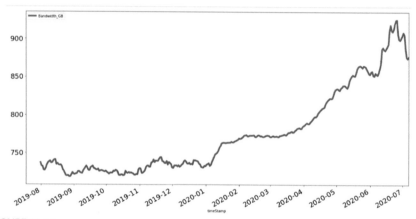

FIGURE 13.12
Removing the seasonality.

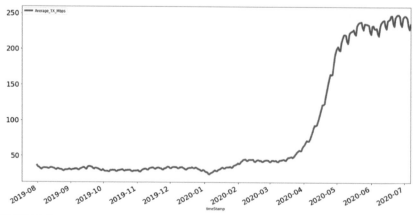

FIGURE 13.13
Visualize under TX speeds.

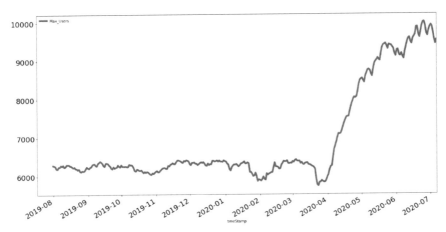

FIGURE 13.14
Maximum number of users.

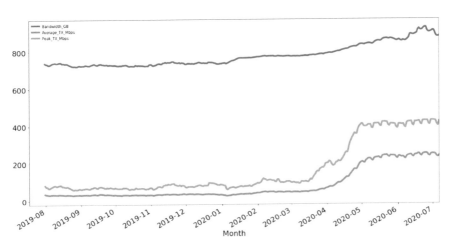

FIGURE 13.15
Upward trends of individual network from April 2020.

To compare the trends of individual network parameters, all the rolling averages are plotted in a single graph. It is observed that there is an upward trend in all the parameters from April 2020 onwards.

13.4.4   Time Series Data: Seasonal Patterns

To analyze seasonality more readily, remove the trend from a time series. This is one approach to considering the seasonal elements of your day-time TA's series. Subtract the original signal from the trend you calculated earlier to get rid of the trend (rolling mean). However, the quantity of data points you averaged will influence this data.

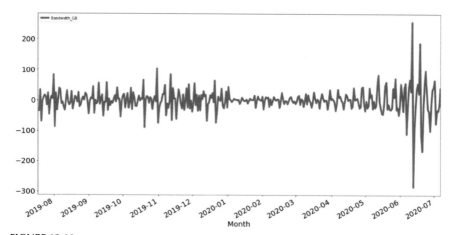

FIGURE 13.16
First order difference of the 'bandwidth' data series.

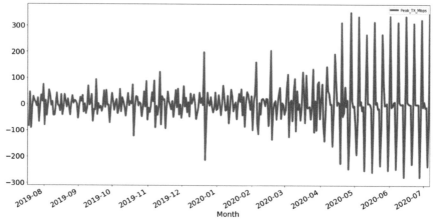

FIGURE 13.17
First order differencing plots of TX speeds.

Another method for removing the trend is "differencing," which involves comparing the differences between successive data points as shown in Figure 13.16 (known as "first-order differencing" because you're simply comparing one data point to the one before it).

Similarly, first-order differencing plots of TX speeds shown in Figure 13.17 and number of users shown in Figure 13.18 show certain seasonability in the data.

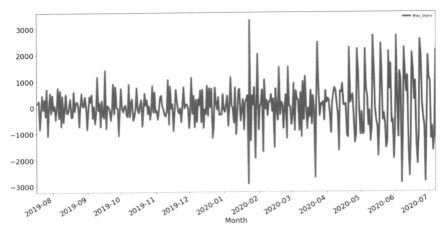

FIGURE 13.18
First order differencing plots of maximum users.

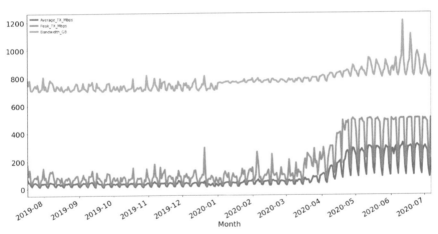

FIGURE 13.19
First order differencing of all network parameters.

All the time series variables are plotted again on the same graph to see how they look like. The first-order differencing of all network parameters is plotted in one graph as in Figure 13.19 to recheck the seasonal pattern if any. From the above visualization, certain amount of seasonability in the time series data for all the variables can be observed as in Figure 13.20 and also an upward trend from the second quarter of 2020 onward.

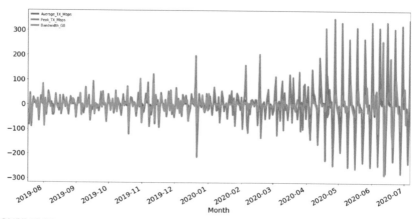

FIGURE 13.20

Seasonability in the time series data for all the variables.

13.5   Implementing the ARIMA Model

Splitting the data into train and test sets

```
1   train
```

time Stamp	Average_TX_Mbps	Peak_TX_Mbps	Max_Users	Bandwidth_GB
2019-07-17 15:00:00	71.75	172.66	6509	782.06
2019-07-18 15:00:00	47.06	90.46	6661	747.22
2019-07-19 15:00:00	44.98	135.63	6895	781.23
2019-07-20 15:00:00	27.96	45.54	6052	711.90
2019-07-21 15:00:00	17.48	36.36	5864	707.87
...	...	...	...	...
2020-06-26 15:00:00	294.92	501.81	10659	918.17
2020-06-27 15:00:00	138.53	224.61	7803	818.29
2020-06-28 15:00:00	98.63	180.00	7293	808.19
2020-06-29 15:00:00	263.09	502.08	9231	845.28
2020-06-30 15:00:00	290.02	490.94	10302	887.56

350 rows × 4 columns

Training data

```
1   test
```

time Stamp	Average_TX_Mbps	Peak_TX_Mbps	Max_Users	Bandwidth_GB
2020-07-01 15:00:00	301.74	500.26	11271	954.16
2020-07-02 15:00:00	281.16	486.70	10047	878.10
2020-07-03 15:00:00	265.37	475.48	9282	846.60
2020-07-04 15:00:00	126.85	239.68	7599	812.17
2020-07-05 15:00:00	79.00	158.53	6426	795.02
2020-07-06 15:00:00	201.92	501.91	8568	835.48

Testing data

13.5.1 Building ARIMA Model

The model building took 8.62 seconds.

```
1  model.summary()
```

Statespace Model Results

Dep. Variable:			y	No. Observations:		356
Model:	SARIMAX(0, 1, 1)x(1, 0, 2, 12)			Log Likelihood		-1764.644
Date:		Wed, 29 Jul 2020		AIC		3541.289
Time:		19:14:08		BIC		3564.522
Sample:			0	HQIC		3550.532
			- 356			
Covariance Type:			opg			

	coef	std err	z	P>|z|	[0.025	0.975]
intercept	0.6396	0.325	1.968	0.049	0.003	1.276
ma.L1	-0.8914	0.019	-45.751	0.000	-0.930	-0.853
ar.S.L12	-0.4642	0.162	-2.858	0.004	-0.783	-0.146
ma.S.L12	0.4667	0.173	2.698	0.007	0.128	0.806
ma.S.L24	-0.3223	0.072	-4.448	0.000	-0.464	-0.180
sigma2	1193.7937	32.073	37.221	0.000	1130.932	1256.656

Ljung-Box (Q):	541.52	Jarque-Bera (JB):	10733.44
Prob(Q):	0.00	Prob(JB):	0.00
Heteroskedasticity (H):	3.98	Skew:	3.06
Prob(H) (two-sided):	0.00	Kurtosis:	29.23

```
1  model.fit(train['Bandwidth_GB'])
```

```
ARIMA(callback=None, disp=0, maxiter=None, method=None, order=(0, 1, 1),
      out_of_sample_size=0, scoring='mse', scoring_args={},
      seasonal_order=(1, 0, 2, 12), solver='lbfgs', start_params=None,
      suppress_warnings=True, transparams=True, trend=None,
      with_intercept=True)
```

For forecasting the next six values of bandwidth, we have six observations in data set. We will then map these forecasts to start index of the test data for comparison.

13.5.2 Printing the Forecasted Values of Bandwidth

```
1  forecast_df
```

| | Prediction |
time Stamp	
2020-07-01 15:00:00	895.349462
2020-07-02 15:00:00	910.456485
2020-07-03 15:00:00	898.725879
2020-07-04 15:00:00	877.606389
2020-07-05 15:00:00	777.374204
2020-07-06 15:00:00	857.398884

Let us concatenate the forecasted data frame with original data frame and plot it (Figure 13.21).

Using the ARIMA model that was built, the prediction of the bandwidth requirement of next two months is done and checked if the predictions are in line with current spike in network utilization due to COVID situation. The prediction of plot of this two-month forecast along with original data set is done to calculate the bandwidth required in August and September 2020 as shown in Figure 13.22. It is also observed the range of forecast values show the interval within which the predictions will fall for future dates. As we can see, the bandwidth requirement will remain in the range of 800 to 1000 during August and September 2020 period as per the time series model as in Figure 13.23.

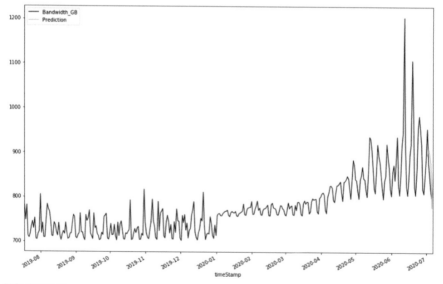

FIGURE 13.21

The predictions are in line with high bandwidth utilization in 2020.

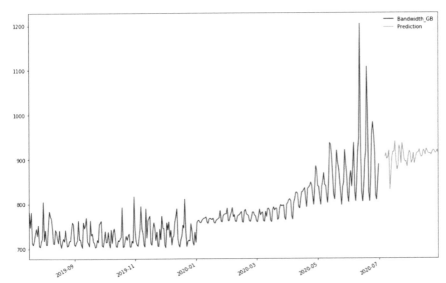

FIGURE 13.22
Bandwidth required in August and September 2020.

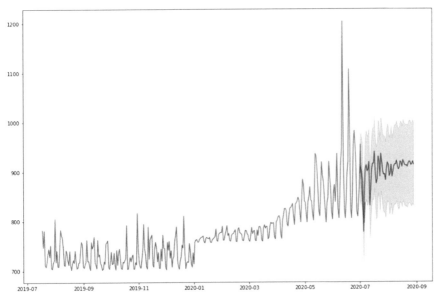

FIGURE 13.23
Bandwidth range of 800–1000 during August and September 2020.

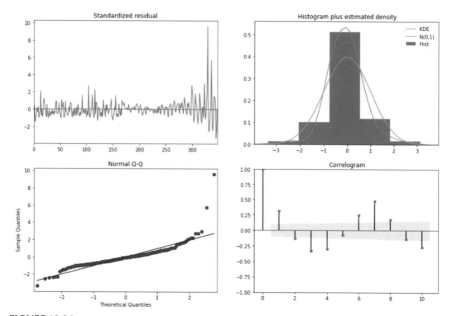

FIGURE 13.24

Evaluation of this proposed model using the diagnostic plot.

13.5.3   Model Evaluation

The evaluation of this proposed model can be done using the diagnostic plot is shown in Figure 13.24.

Standardized Residuals: The residual errors appear to have a uniform variance and fluctuate around a mean of zero. The density plot indicates a normal distribution with a mean of zero.

Normality: The dots on the Q–Q plot should all be completely aligned with the red line. Any large variances would indicate a skewed distribution.

The ACF plot, also known as the correlogram, illustrates that the residual errors are not autocorrelated. Any autocorrelation would imply that the residual errors have a pattern that isn't explained by the model. As a result, you'll need to add more Xs (predictors) to the model. In general, it appears to be a decent match.

13.6   Conclusion: Business Benefits of the Models

13.6.1   Network Traffic Classification

Network traffic categorization is a method that divides computer network traffic into different classifications based on various criteria (such as port number or protocol). Using classified network traffic, you may do tasks like monitoring, discovery, control, and optimization. The overall purpose of network traffic classification is to improve network performance.

The packets can be labelled once they have been classified as belonging to a specific application. These markings or flags will assist the network router in determining the proper service policies for certain network flows.

This study's classification approach can be used in conjunction with network or packet monitoring software. This allows us to create a real-time dashboard that shows application-specific network use in real time. This will aid in the rapid and efficient regulation of traffic flow across any internet link.

13.6.2 Time Series Prediction

Due to the introduction of new technology, industries, and applications, internet traffic has increased dramatically during the last decade. The Internet of Things (IoT), Cloud computing, and data-center applications are all growing in popularity. Networks must cope with ever-increasing traffic demands while maintaining a high level of service to users. As a result, effective networking equipment usage has become critical. If network traffic can be properly forecast, this task can be completed effectively. Many networking sectors, such as energy conservation, network resource management, and wireless sensor networks, can benefit from accurate traffic forecast.

One advantage of traffic prediction is that it saves energy. Traffic prediction can be utilized in internet core network routers/switches to save a large amount of power. The number and sophistication of processors employed in these network devices are expanding in response to rising traffic demands and computing requirements, resulting in increased power consumption. High equipment power consumption and accompanying increases in cooling expenses result in increased network operational costs. If network traffic can be precisely forecast, additional processors in these core devices can be turned down to save power during low traffic times. Construction of greener traffic-aware networks can accurately estimate on future traffic loads.

Second, in order to make an optimum use of network resources, traffic forecast is essential. Accurate network traffic prediction at access points provides efficient resource allocation to ensure acceptable service quality.

Finally, when new applications such as e-commerce and online banking emerge, identifying and preventing network abuse is becoming increasingly challenging due to increased traffic and network complexity. Anomaly detection is one method of preventing network abuse. An assault can be detected by a considerable divergence from typical traffic behavior. The accuracy of traffic forecast is directly proportional to the performance of anomaly detection.

Bibliography

[1] Shahraki, Amin, Mahmoud Abbasi, Amir Taherkordi, and Anca Delia Jurcut. "Active learning for network traffic classification: A technical survey." *arXiv preprint* arXiv:2106.06933 (2021).

[2] Callegari, Christian, Pietro Ducange, Michela Fazzolari, and Massimo Vecchio. "Explainable Internet Traffic Classification." *Applied Sciences* 11, no. 10 (2021): 4697.

[3] Abbasi, Mahmoud, Amin Shahraki, and Amir Taherkordi. "Deep learning for network traffic monitoring and analysis (NTMA): A survey." *Computer Communications* 170 (2021): 19–41.

[4] Dong, Shi. "Multi class SVM algorithm with active learning for network traffic classification." *Expert Systems with Applications* 176 (2021): 114885.

[5] Zhongsheng, Wang, Wang Jianguo, Yang Sen, and Gao Jiaqiong. "Traffic identification and traffic analysis based on support vector machine." *Concurrency and Computation: Practice and Experience* 32, no. 2 (2020): e5292.

[6] Salman, Ola, Imad H. Elhajj, Ayman Kayssi, and Ali Chehab. "A review on machine learning-based approaches for internet traffic classification." *Annals of Telecommunications* 75, no. 11 (2020): 673–710.

[7] Network Based Application Recognition (NBAR)—Cisco, 2009: www.cisco.com/c/en/us/products/ios-nx-os-software/network-based-application-recognition-nbar/index.html. (Accessed March 12, 2021)

[8] Palo Alto Networks, Next-Generation Firewall Overview, 2011.

[9] www.paloaltonetworks.com/content/dam/paloaltonetworks-com/en_US/assets/pdf/datasheets/firewall-features-overview/firewall-features-overview.pdf. (Accessed March 15, 2021)

[10] Mellia, Marco, Antonio Pescapè, and Luca Salgarelli. "Traffic classification and its application to modern networks." *Computer Networks* (1999) 53, no. 6 (2009).

[11] Risso, Fulvio, Mario Baldi, Olivier Morandi, Andrea Baldini, and Pere Monclus. "Lightweight, payload-based traffic classification: An experimental evaluation." In *2008 IEEE International Conference on Communications*, pp. 5869–5875. IEEE, 2008.

[12] Avalle, Matteo, Fulvio Risso, and Riccardo Sisto. "Efficient multistriding of large non-deterministic finite state automata for deep packet inspection." In *2012 IEEE International Conference on Communications (ICC)*, pp. 1079–1084. IEEE, 2012.

[13] Crotti, Manuel, Maurizio Dusi, Francesco Gringoli, and Luca Salgarelli. "Traffic classification through simple statistical fingerprinting." *ACM SIGCOMM Computer Communication Review* 37, no. 1 (2007): 5–16.

[14] Bujlow, Tomasz, Tahir Riaz, and Jens Myrup Pedersen. "A method for classification of network traffic based on C5. 0 Machine Learning Algorithm." In *2012 International Conference on Computing, Networking and Communications (ICNC)*, pp. 237–241. IEEE, 2012.

[15] Nguyen, Thuy T.T., and Grenville Armitage. "A survey of techniques for internet traffic classification using machine learning." *IEEE Communications Surveys & Tutorials* 10, no. 4 (2008): 56–76.

[16] Ubik, Sven, and Petr Žejdl. "Evaluating application-layer classification using a machine learning technique over different high-speed networks." In *2010 Fifth International Conference on Systems and Networks Communications*, pp. 387–391. IEEE, 2010.

[17] Bujlow, Tomasz, and Jens Myrup Pedersen. "Multilevel classification and accounting of traffic in computer networks." *Classification and Analysis of Computer Network Traffic* 167.

[18] Bujlow, Tomasz, and Jens Myrup Pedersen. "A practical method for multi-level classification and accounting of traffic in computer networks" Aalborg Universitet (2014).

[19] Jun, Li, Zhang Shunyi, Lu Yanqing, and Zhang Zailong. "Internet traffic classification using machine learning." In *2007 Second International Conference on Communications and Networking in China*, pp. 239–243. IEEE, 2007.

[20] Bujlow, Tomasz, and Valentin Carela-Espanol. "Comparison of deep packet inspection (DPI) tools for traffic classification" (UPC-DAC-RR-CBA-2013-3 ed.) Universitat Politècnica de Catalunya. www.ac.upc.edu/app/research-reports/html/research_center_index-CBA-2013,en.html (2013).

[21] Dusi, Maurizio, Francesco Gringoli, and Luca Salgarelli. "Quantifying the accuracy of the ground truth associated with Internet traffic traces." *Computer Networks* 55, no. 5 (2011): 1158–1167.

[22] Cascarano, Niccolò, Luigi Ciminiera, and Fulvio Risso. "Optimizing deep packet inspection for high-speed traffic analysis." *Journal of Network and Systems Management* 19, no. 1 (2011): 7–31.

[23] Alcock, Shane, and Richard Nelson. "Measuring the accuracy of open-source payload-based traffic classifiers using popular internet applications." In *38th Annual IEEE Conference on Local Computer Networks-Workshops*, pp. 956–963. IEEE, 2013.

[24] Goss, Ryan, and Reinhardt Botha. "Deep packet inspection—Fear of the unknown." In *2010 Information Security for South Africa*, pp. 1–5. IEEE, 2010.

[25] Valenti, Silvio, Dario Rossi, Alberto Dainotti, Antonio Pescapè, Alessandro Finamore, and Marco Mellia. "Reviewing traffic classification." In *Data Traffic Monitoring and Analysis*, pp. 123–147. Springer, Berlin, Heidelberg, 2013.

[26] Wei, Yong, Z. Yun-Feng, and Li-chao Guo. "Analysis of message identification for OpenDPI." *Computer Engineering* (2011): S1.

[27] Alcock, Shane, and Richard Nelson. *Libprotoident: Traffic Classification Using Lightweight Packet Inspection*. Technical report, University of Waikato, 2012.

[28] Levandoski, Justin. "Application layer packet classifier for Linux." http://l7-filter. sourceforge. net/ (2008).

[29] Dainotti, Alberto, Walter De Donato, and Antonio Pescapé. "Tie: A community-oriented traffic classification platform." In *International Workshop on Traffic Monitoring and Analysis*, pp. 64–74. Springer, Berlin, Heidelberg, 2009.

[30] Khakpour, Amir R., and Alex X. Liu. "High-speed flow nature identification." In *2009 29th IEEE International Conference on Distributed Computing Systems*, pp. 510–517. IEEE, 2009.

[31] Kim, Hyunchul, Kimberly C. Claffy, Marina Fomenkov, Dhiman Barman, Michalis Faloutsos, and KiYoung Lee. "Internet traffic classification demystified: myths, caveats, and the best practices." In *Proceedings of the 2008 ACM CoNEXT conference*, pp. 1–12. 2008.

[32] Aceto, Giuseppe, Alberto Dainotti, Walter De Donato, and Antonio Pescapé. "Portload: taking the best of two worlds in traffic classification." In *2010 INFOCOM IEEE Conference on Computer Communications Workshops*, pp. 1–5. IEEE, 2010.

[33] Kumar, Sailesh, Sarang Dharmapurikar, Fang Yu, Patrick Crowley, and Jonathan Turner. "Algorithms to accelerate multiple regular expressions matching for deep packet inspection." *ACM SIGCOMM Computer Communication Review* 36, no. 4 (2006): 339–350.

[34] Cascarano, Niccolo, Pierluigi Rolando, Fulvio Risso, and Riccardo Sisto. "iNFAnt: NFA pattern matching on GPGPU devices." *ACM SIGCOMM Computer Communication Review* 40, no. 5 (2010): 20–26.

[35] Bujlow, Tomasz, Valentín Carela-Español, and Pere Barlet-Ros. "Extended independent comparison of popular deep packet inspection (DPI) tools for traffic classification." (2014). Universitat Politècnica de Catalunya. www.ac.upc.edu/app/research-reports/html/research_center_index-CBA-2014,en.html

[36] Fusco, Francesco, and Luca Deri. "High speed network traffic analysis with commodity multi-core systems." In *Proceedings of the 10th ACM SIGCOMM Conference on Internet Measurement*, pp. 218–224. 2010.

14

Integration of AI/Ml in 5G Technology toward Intelligent Connectivity, Security, and Challenges

Devasis Pradhan,[1,*] **Prasanna Kumar Sahu,**[2] **Rajeswari,**[1] **Hla Myo Tun,**[3] **and Naw Khu Say Wah**[3]

[1]*Department of Electronics and Communication Engineering, Acharya Institute of Technology, Bangalore*

[2]*Department of Electrical Engineering, National Institute of Technology, Rourkela*

[3]*Department of Electronic Engineering, Yangon Technological University, Yangon, Myanmar*

[*]*devasispradhan@acharya.ac.in*

CONTENTS

DOI: 10.1201/9781003104858-14

14.1 Introduction

In the ceaselessly developing correspondence organization engineering, to coordinate different scopes of gadgets with remarkable prerequisites for an alternate organization, parameters have brought about refined difficulties for network security. The new improvements in 5G networks and past are working with the vivid development of information correspondence by giving higher information rates. 5G is a versatile network confronting many difficulties to meet the remarkable developing requests for admittance to remote administrations with super low idleness and, further, high information rates. 5G today is the central innovation of many state-of-the-art advancements like the IoT, smart grid, automated aeronautical frameworks, and self-driving vehicles. 5G is needed to be described by high adaptability in plan and assets by the executives and allotted to fulfill the expanding needs of this heterogeneous network and clients or end-users[1–3].

Numerous wireless examinations driving gatherings foresee that artificial intelligence (AI) as the next huge "game-evolving" innovation, ready to furnish 5G with the adaptability and the knowledge required. Hence numerous analysts have examined the proficiency of this hypothesis in numerous parts of 5G remote interchanges including regulation, channel coding, obstruction of the executives, planning, 5G slicing, reserving, energy productivity, and digital protection. ML and AI algorithms can be utilized to process and dissect cross-space information that would be needed in 5G in a considerably more effective manner empowering speedy choices and as such facilitating the organization's intricacy and decreasing the support cost. The cross-space information incorporates geographic data, designing boundaries, and different information to be utilized by AI and ML to all the more likely gauge the pinnacle traffic, streamline the organization for limit extension, and empower more canny inclusion through powerful impedance estimations [4].

14.2 Overview of 5G

The quick reception of 5G innovation is promising a stunning number of new gadgets. For instance, the Cisco Annual Internet Report (2018–2023) estimates that "Machine-to-Machine (M2M) communication will develop 2.4 overlays, from 6.1 billion out of 2018 to 14.7 billion by 2023. There will be 1.8 M2M communication for every individual from the worldwide populace by 2023." The outstanding development in associated gadgets alongside the presentation of 5G innovation is relied upon to cause a test for productive and dependable organization's asset assignment. In addition, the gigantic organization of the

Internet of Things and associated gadgets to the internet might pose a genuine danger to the organization's security assuming they are not taken care of appropriately [5,6]. The 5G networks are relied upon to help a lot higher level heterogeneity (as far as associated gadgets and networks) when contrasted with their archetypes. For example, 5G supports intelligent vehicles, shrewd homes, savvy structures, and shrewd urban areas. Also, the IoT in the 5G design will include more strong and versatile strategies to deal with the basic security issues both at the organization and gadget sides. The security of such a network will be considerably more convoluted due to the external interruption just as the nearby interruption. Artificial intelligence and ML can give arrangements by grouping delicate security connections in the middle, for example, character, validation, and confirmation. The security and security in 5G-IoT will cover every one of the layers like character assurance, security, and protecting end-to-end users. For example, the key validation system from end-gadget to core network and onward to the specialist co-op, while hiding the key identifier, is as yet a mind-boggling issue. AI and ML can likewise assume a significant part in key validation alongside adequately limiting the disguising attacks.

14.2.1 Enabling Technologies

5G will empower its availability billions of gadgets in the IoT and the associated world. Thus, there are three significant classifications of utilization cases for 5G.

(a) *Enormous machine to machine communication*: It's additionally considered IoT, which includes associating billions of gadgets without human intercession at a scale not seen previously. This has cutting-edge modern interaction and applications including agribusiness, producing, and business correspondence.

(b) *Low latency with ultra-reliability*: It incorporates ongoing control of gadgets, modern advanced mechanics, vehicle-to-vehicle correspondence likewise opening up another reality where far-off clinical consideration, strategy, and treatment are conceivable.

(c) *Improved versatile broadband*: It gives basically quicker information speed and more prominent limit keeping the world associated. The clever application will incorporate fixed remote Web access for homes and outside broadcast application networks for individuals moving.

14.2.2 Key Enablers for 5G

The key enabling techniques have been elaborated in Table 14.1 with their applicability. These key enablers help the 5G network to make more cognitive and versatile application-oriented networks.

TABLE 14.1

Key Enablers for 5G Network

S. No.	Techniques	Applicability	Support to Network	Challenges
1	NFV (network function virtualization)	Asset usage further developed throughput, energy investment funds, decreased CAPEX/OPEX, upgraded QoE, more straightforward movement/support.	Data-centric communication (everything as assistance)	Disengagement, asset designation, decency, income/value improvement, versatility the executives
2	C-RAN (Cloud-RAN)	Adaptability, energy/power investment funds, expanded throughput, decreased deferral, flexibility to dynamic traffic, diminished CAPEX/OPEX, simpler management of used network	Administration situated communication with heterogeneous network (HetNet)	BBU managements (e.g., participation, interconnection, bunching), energy-mindful planning, fronthaul-mindful asset designation
3	HetNet (Heterogeneous Network)	Expanded throughput, range use, energy productivity, inclusion extension	Small cell network, D2D/M2M communication, Internet-of-Things	Mitigating the interference versatile power control, dynamic mode choice and offloading to underlay organization, gadget revelation, bound together MAC plan
4	FDC (Full Duplex Communication)	Range productivity, decreased inertness, energy effectiveness	Small cell network, D2D/M2M, intellectual radio networks, multihop transferring	SI decrease, cross-layer asset the executives, power portion, impedance the board, synchronization and time change in accordance with setting up FD transmission, dynamic mode determination, planning a MAC convention
5	RF-EH (RF Energy Harvesting)	Green communication with energy efficiency	Small cell network, D2D/M2M communication,	Muti-client planning, progressed channel securing, energy beamforming, gather/communicate time variation, obstruction the board, SWIPT- empowered asset distribution

TABLE 14.2

5G Requirements as Compared to 4G

S. No.	Verticals	Remarks
1	Speed	1–10 Gbps associations with end focuses in the field (for example, not hypothetical greatest)
2	Latency	1 millisecond start to finish full circle delay—dormancy
3	BW utilization	1000× transfer speed per unit region
4	Device to be connected	10–100× number of associated devices
5	Accessibility	Perception of 99.999% accessibility
6	Coverage	100%
7	Usage of energy	90% decrease in network energy use
8	Battery life of device	As long as ten-year battery life for low power, machine-type gadgets

14.2.3 5G Requirements

Lately, there have been a few perspectives about a definitive structure that 5G technology should take. There have been two perspectives on what 5G ought to be:

(a) *Hyper-connectivity*: This perspective on the prerequisites for 5G remote frameworks means to take the current innovations including 2G, 3G, 4G, Wi-Fi, and other significant remote frameworks to give higher inclusion and accessibility, alongside more thick organizations. Aside from having necessities to offer conventional types of assistance, a key differentiation is empowered new administrations like machine to machine (M2M) applications alongside extra Internet of Things (IoT) applications [7]. This arrangement of 5G necessities could require another radio innovation to empower low power, low throughput field gadgets with long battery lifetimes of ten years or more.

(b) *Radio-Access Technology* (RAT): This perspective on the 5G prerequisites takes a more innovation-driven view and sets details for information rates, inertness, and other key boundaries. These necessities for 5G would empower an unmistakable boundary to be made between 4G or different administrations and the new 5G remote framework [8, 9].

14.3 Artificial Intelligence (AI) and Machine Learning

The idea of utilizing AI and ML in security isn't new, yet their attainability and execution prevalence generated interest with the advancement of deep learning (DL) calculations. A large portion of the strategies previously aimed at the improvement of DL was devoted to displaying the assault designs with specific attributes that are not vigorous in nature, yet with AI and ML, it is normal that frameworks will turn out to be stronger toward new refined dangers and assaults with dynamic attributes. This is because aggressors utilize complex methods like confusion, polymorphism, or pantomime to stay away from recognition. From bundle catching and examination to enormous information bits of knowledge, AI and ML can be utilized to advise the dangers not distinguished by ordinary methods. The example-based learning at the center is upheld by softwarization furthermore, and virtualization gives spryness and vigor to convenient counter cyber attacks [10–12].

Artificial intelligence calculations are being used for security and protection issues. The data security industry is creating an ever-increasing number of information that opens them to propel threats and AI could be an amazing antitoxin. The original AI arrangements are zeroing in on examining information, recognizing the threat, and helping people in the remediation plan. The second era of AI will make the frameworks more independent and just leave the basic help issues to people [10, 11].

14.3.1 Amalgam of AI and ML in 5G

An expanded transfer speed, higher range use, and further, high information rates in 5G organizations have additionally enlarged the threat and security scene from individual gadgets to the specialist co-op network. Consequently, the network ought to be savvy enough to manage these difficulties continuously and ML and AI methods could assist with demonstrating these strong unique calculations in recognizing network issues and furnish with the conceivable arrangement progressively. In short to medium-term plans, AI and ML can be utilized to distinguish the dangers and counter them with hearty and versatile security calculations [13]. Though, in the long haul, a completely mechanized security instrument is imagined for opportune reaction to threats and cyberattacks.

The 5G network relied upon to help a lot higher level heterogeneity (as far as associated gadgets and networks) when contrasted with its archetypes. For example, 5G supports brilliant vehicles, shrewd homes, savvy structures, and smart urban areas. Additionally, the IoT in 5G network design will include more hearty and versatile strategies to deal with the basic security issues both at the network and gadget sides [13, 14].

14.3.2 Security and Privacy

The security of such networks will be significantly more confounded as a result of the external interruption just as the nearby interruption. AI and ML can give arrangements by arranging delicate security connections in the middle, for example, identity, confirmation, and protecting the user details. The security and privacy in 5G-IoT will cover every one of the layers like personality insurance, protection, and E2E protection. Catering for security and protection of information from these various frameworks with extraordinarily unique security prerequisites becomes a monotonous undertaking [15]. AI and ML with an outline of SBA and security prerequisites for distinctive end-frameworks can identify and redress these issues progressively by arranging and bunching surprising dangers. AI and ML can help in creating security instruments by making trust models, gadget security, and information confirmation to give precise security for the entire 5G-IoT network [16,17].

TABLE 14.3

AI Techniques for 5G

S. No.	AI Learning Scheme	Leaning or Training Model	Application
1	Supervised	ML and statistical logistic	Dynamic recurrence and transfer speed portion in self-coordinated LTE dense little cell arrangements. Path loss
		SVM (Support Vector Machine)	Prediction of Pathloss estimation
		Neural Network Based	Channel learning to derive undetectable channel state information (CSI) from a recognizable channel
		ANN and MLPs	Demonstrating and approximations of true capacities for interface spending plan and proliferation misfortune for cutting edge wireless network.
2	Unsupervised	Gaussian Mixture Model (GMM)	Cooperative spectrum sensing for vehicular networks
		Hierarchical clustering (HCM)	Fault detection
		Affinity clustering model (ACM)	Resource management for ultra-small cell network
3	Reinforcement	Long short-term memory model	Proactive asset allocation in LTE-U Networks, formed as a non-cooperative game, which empowers SBSs to realize which unlicensed channel, given the drawn-out WLAN movement in the channels and LTE-U traffic loads
		Grad follower (GF) and MBM	Empower femto-cells (FCs) to independently and astutely detect the radio climate and tune their boundaries in HetNets, to decrease intra/inter-level impedance

14.3.3 AI Learning-Based Scheme for 5G

(a) *Supervised learning scheme*: In supervised learning, each preparation model must be taken care of alongside its individual name. The thought is that preparation of a learning model is an example of the issue occasions with known optima, and afterward, utilize the model to perceive ideal answers for new cases. An average undertaking on directed learning is to foresee an objective numeric worth, given a bunch of highlights, called indicators. This depiction of the undertaking is called relapse. Move learning is a famous procedure frequently used to group vectors. Basically, one would prepare a convolutional neural network (CNN) on an extremely huge data set, for instance, Image Net [14], and afterward adjust the CNN on an alternate vector data set. The great part here is preparing on the huge data set is as of now done by certain individuals who offer the learned loads for public examination use.[16–18]

(b) *Unsupervised Learning Scheme*: In unsupervised learning, the preparation information is unlabeled, and the framework endeavors to learn with no direction. This strategy is especially helpful when we need to distinguish gatherings of comparative attributes. Never do we advise the calculation to attempt to recognize gatherings of related characteristics; the calculation tackles this association without intercession [16,17].

(c) *Reinforcement Learning Scheme*: It depends on a learning framework frequently called a specialist, which responds to the climate. The specialist performs activities and gets prizes or punishments (negative prizes) as a trade-off for its activities. That implies that the specialist needs to learn without help from anyone else making an approach that characterizes the activity that the specialist ought to pick in a specific circumstance. The point of the support learning task is to expand the previously mentioned award over the long run [16,17].

14.4 Wireless AI

5G communication and network management is enabled by machine learning and deep learning. For every model, the benefits and weaknesses of AI-empowered procedures is discussed as follows:

(a) *M-MIMO and beamforming*: Massive MIMO is one element of 5G. Using countless radiating antennas, 5G can concentrate the transmission and gathering of signal power into small regions. Be that as it may, a few issues are identified with this innovation. AI/ML has been applied in massive MIMO to beat these issues. For example, a precise gauge of the channel with basic assessment strategies and a sensible number of pilots is tried in huge MIMO: the low intricacy least-squares (LS) assessor doesn't accomplish an acceptable execution, while least mean square error (MMSE) channel assessment is extremely intricate. AI and deep learning have been likewise examined in enhancing the loads of receiving wire components in enormous MIMO. Profound learning and AI can foresee the client appropriation and in like manner improve loads of antenna components, which can work on the inclusion in a multi-cell situation [5, 19].

(b) *Channel coding*: An observable component of the air interface of the 5G is the utilization of new channel coding methods: Data channels utilize low-thickness equality check (LDPC) codes, and control channels utilize polar codes [8]. For example, polar codes can accomplish astounding execution; however, it takes a few emphases to accomplish this exhibition, and it is absolutely impossible to anticipate how quickly polar codes can arrive at this ideal presentation. Profound learning is notable for its high parallelism structure, which can carry out a single shot coding/deciphering. Subsequently, numerous specialists foresee that profound learning-based channel coding is a hopeful technique to empower 5G NR. The DL-based channel coding can accomplish a decent scope of execution intricacy compromises, in case the preparation is performed effectively as the decision of code-word length causes over-fitting and under-fitting [8, 9, 20].

(c) *Energy-efficient network*: The usage of ICT is answerable for 2–10 percent of the world's energy utilization in 2007, and it is relied upon to for further development [21]. Additionally, over 80 percent of ICT is from a radio access network (RAN), which is sent to meet the pinnacle traffic burden and stays on it regardless of whether the heap is light. This goal can be accomplished by lessening the power utilization of the base stations and cell phones. AI/ML adapting along these lines can help in building clever remote organizations that proactively foresee the traffic and versatility of clients and conveyance benefits just when mentioned, therefore, decreasing the power utilization in radio access networks (RAN).[7–9, 20]

(d) *5G network slicing*: Two conspicuous elements of 5G are network slicing and caching. The first permits administrators to convey diverse assistance types in more than one network framework. The last option predicts the substance that clients might demand productive use of the capacity of the base station. 5G asset provisioning and storing utilize the hypothesis of AI/ ML [22].

(e) *Cognitive radio*: Radio access is scant, and there is an expanding interest in remote traffic. Insightful remote organization executives is the way forward to meet these expanding requests. AI/ML can be a promising to be included for asset assignment in 5G networks. DL can be a decent option for the avoidance of interference in the network, usage of spectrum, multi-path utilization with less path loss, interface variation, multi-channel access, and traffic blockage [23].

(f) *Modulation*: AMC is a central strategy in non-cooperative communication frameworks. Modulation regulation is one errand that can help in arranging the balance sort of a got signal, which is an important stage toward comprehension and detecting the remote climate. Great detecting and variation work on ghastly proficiency and obstruction alleviation. Profound learning-based AMC frameworks comprise three parts: The initial segment is signal handling to upgrade the nature of the got tests, recurrence offset rectification, gain control, enhancers, and separating. The subsequent part includes the extraction of elements like the abundance, stage, and recurrence of the got signal. The last part is a sign classifier: characterization of the balance types. AI/ML can accomplish the high exactness of adjustment characterization [23,24].

14.5　AI in Context with 5G

With regards to 5G and the improvement of versatile organizations, AI and ML are reciprocally utilized, yet they contrast from one another. AI is an expansive idea that does specific undertakings astutely and nearer to people. It depends on ML to gather the information and investigate the example from which the product framework learns and improves, and this makes machines more intelligent. On the opposite side, ML is a subset of AI and is viewed as a use of AI to permit machines to access information and assess on their own. ML is otherwise called a present status of the craft of AI [25]. The complexity of 5G can be overcome using two fundamental approaches:

(a) *Basic*: The fundamental methodology where AI/ML is utilized to play out some essential assignments is dependent on some preset calculations or algorithms.

(b) *Advanced*: In this developed methodology, AI/ML is utilized to be more mindful of settings and gain from the encompassing circumstances and acts as:

　(i)　This methodology has arisen with the prevalence of the internet and the enormous measure of the created data.

　(ii)　Instead of training the PC to do everything, it is smarter to code them to think like people and give them admittance to the tremendous data empowered by the Internet and 5G.

14.6 Intelligent Connectivity

It is a new idea, which is based on a combination of three significant innovations, 5G and IoT, which is intended to fill in as a way to speed up the advancement of troublesome computerized administrations. This new idea works with the association of gadgets through a quick and low inactivity portable organization, that is 5G, which gathers advanced data through the machines and sensors, which is the capacity of IoT. At that point, examinations and contextualizations by AI/ML lastly creates significant results that are helpful for the clients [25,26]. This would empower new groundbreaking capacities in the greater part of the business areas, for example transport, producing, medical care, public well-being, security, and so forth.

(a) *Planning of network*: The operator will utilize AI to further develop the organizational scope quantification, which will prompt a decrease in expenses and better execution of the organization. Artificial intelligence and ML can be applied to foresee and conjecture traffic by distinguishing traffic designs and as such learning on the web and assisting with the mechanization of choices. This voids the requirement for over-arrangement just like the case with the conventional organization limit arranging [27].

(b) *Performance of network to be monitored*: This permits a network regulator to gain as a matter of fact while it upgrades the network.

(c) *Customer satisfaction*: AI will assist with improving and dealing with the client experience by utilizing the IoT information that uncovers significant shopper bits of knowledge with regards to real-time circumstances. This aids the purchasers and furnishes them with an encounter tailor-made to their life [27,28].

(d) *Product life cycle*: AI/ML assists with overseeing and further developing the item life cycle by utilizing the information that portrays the current and recorded item experiences, main drivers and relationships, future results, and suggested upgrades. This assists with changing the item life cycle from an information in the executives instrument to an insightful dynamic framework toward cognitive connectivity [28,29]

(e) *Revenue generation*: The integration of AI/ML in 5G is as of now occurring and will lessen the capital consumption and as such further develop the income stream as well as to utilize the devices more effectively with less utilization of energy [29,30].

14.7 5G Security

Generally, AI and ML calculations are information hungry in nature, which implies that information is expected to prepare the model for powerful working. In the period of 5G, information age, stockpiling, and the management of data are quite easy as we have high computational power, outstanding information development, and information sources. The network can be kept up with, accessed, and examined for potential threats, attacks, and weaknesses utilizing AI and ML at a lower cost of figuring and a reasonable framework [21, 30–32].

AI and ML models can be utilized to recognize dubious exercises continuously by dissecting the designs and boundaries of network movement. Order calculations can be utilized to distinguish inconsistencies by observing organization boundaries such as throughput and organization blunder logs. Grouping algorithms can be utilized to classify different sorts of dangers and also escape clauses in network security. The models, for example, measure induction assaults and generative adversarial networks (GAN) can create counterfeit datasets and these are new safety efforts just like testing and carrying out advanced security conventions and calculations. The exploration in creating private AI and ML models have seen some huge improvement in secure computation, encryption, protection, and unified learning. Half and half models are made by taking on methods from various fields to make models effective, quicker, and summed up [21, 33, 34].

14.8 Challenges

14.8.1 5G Complexity

AI/ML algorithms should be deployed at the time of proper communication in wireless gadgets. Be that as it may, numerous remote gadgets have restricted memory and registering capacities, which isn't appropriate for complex calculations. The assortment of huge examples and preparing profound learning models take extensive time, which is a huge hindrance to convey them on some remote gadgets having restricted power and capacity. Now and again, the higher the quantity of tests and the more huge the preparation time are, the higher the precision of acknowledgment of the sign and organization highlights is. Obtaining more examples and preparing the models for longer occasions bring about lethargic input. Thus, machine learning models ought to be intended to accomplish best precision with less examples and within a brief time frame [34].

14.8.2 Security and Privacy

Safeguarding the security of the clients is the essential worry of portable and specialist co-ops. One of the principal challenges in remote AI is the means by which one can empower the preparation on a data set having a place with clients without sharing the information and putting the individual data of clients in danger. It is important to have a security way to deal with the mix of profound learning in remote interchanges [35]. The security of AI/ML models itself is another test, as the neural network is pruned to antagonistic attacks. Attackers can influence the preparation interaction by infusing counterfeit data sets; such infusion can bring down the precision of the models and yield wrong plan, which may influence the execution in an organization. The examination of ML or AI with regard to security, by and large, remains shallow [35,36].

14.8.3 Trade-off in Speed

The dependability of these procedures is definitely not exactly conventional strategies in remote interchanges in taking care of certain issues. For example, profound learning can contend with LS and MMSE in remote divert assessment in massive MIMO; however, lethargic input portrays these methods. AI/ML induction might lengthen the framework reaction time. This is because not most remote gadgets approach cloud computing, and regardless of the situation, correspondence with cloud servers will present additional postponements.[37–40].

14.9 Research Scope

To facilitate the mix of AI/ML, research endeavors are required in a few ways. For example, the speed increase of a profound neural organization has to be cutting edge for equal registering, quicker calculation, and distributed computing. Appropriated profound learning frameworks present a chance for 5G to assemble the insight in its frameworks to convey high throughput and super low idleness. There have been a few ongoing endeavors in increasing speed of a profound neural organization.

5G-IoT security and protection need more examination in the spaces of confirmation, approval, access control, and protection safeguarding. The current 3GPP-defined networks utilize practical hub detail and conceptual interfaces yet in 5G-IoT. The actual organization will fill in as a center framework and security affirmation will be the vital test to manage. At this stage, semi-managed AI-helped arrangements better suit the appropriate frameworks. With the advancement of AI calculations/algorithms, these frameworks will turn out to be completely mechanized later on.

14.10 Conclusion

In this chapter, we focused on the integration of AI/ML for 5G network with respect to different constraints. We concentrated on a few contextual investigations counting regulation characterization, channel coding, enormous MIMO, reserving, energy proficiency, and network safety. As a conclusion of this top-to-bottom review, AI-empowered 5G communication and systems administration is a promising arrangement that can furnish remote organizations with the insight, effectiveness, and adaptability needed to deal with the alarm radio asset well and convey superior grade of administration to the clients. In any case, a few endeavors are as yet expected to decrease the intricacy of profound adapting so it tends to be executed in time-delicate organizations and low-power gadgets and test the models in more reasonable situations. These days, with the power and pervasiveness of data, various specialists are adjusting their insight and growing their instruments arms stockpile with AI-based models, calculations, and practices, particularly in the 5G world, where even a couple of milliseconds of inactivity can have an effect.

References

[1] C. Zhang, P. Patras, and H. Haddadi, "Deep learning in mobile and wireless networking: A survey," *IEEE Communications Surveys & Tutorials*, 2019.

[2] M. Soltani, V. Pourahmadi, A. Mirzaei, and H. Sheikhzadeh, "Deep learning-based channel estimation," *IEEE Communications Letters*, vol. 23, no. 4, pp. 652–655, 2019.

[3] S. Gao, P. Dong, Z. Pan, and G. Y. Li, "Deep learning based channel estimation for massive MIMO with mixed-resolution ADCS," *IEEE Communications Letters*, vol. 23, no. 11, pp. 1989–1993, 2019.

[4] H. Ye, G. Y. Li, and B.-H. Juang, "Power of deep learning for channel estimation and signal detection in ofdm systems," *IEEE Wireless Communications Letters*, vol. 7, no. 1, pp. 114–117, 2017.

[5] Devasis Pradhan and K.C. Priyanka, "RF-energy harvesting (RF-EH) for sustainable ultra dense green network (SUDGN) in 5G green communication," *Saudi Journal of Engineering and Technology* 2415–6264 (Online, 5(6), pp. 2582013264 DOI: 10.36348/sjet.2020.v05i06.001, 2020)

[6] Devasis Pradhan and K.C. Priyanka, "A comprehensive study of renewable energy management for 5G green communications: Energy saving techniques and its optimization," *Journal of Seybold Report* 1533–9211, 25(10), pp. 270–284. 2020.

[7] Devasis Pradhan and A. Dash, "An overview of beam forming techniques toward the high data rate accessible for 5G networks," *International Journal of Electrical, Electronics and Data Communication*, 2320–2084, 2321–2950, 8(12), pp. 1–5, 2020.

[8] Devasis Pradhan and R. Rajeswari, "5G-green wireless network for communication with efficient utilization of power and cognitiveness," in Jennifer S. Raj (ed.), International Conference on Mobile Computing and Sustainable Informatics. Springer Nature Switzerland AG 2021:Springer, Cham, pp. 325–335. 2020.

[9] Devasis Pradhan, P. K. Sahu, A. Dash, and Hla Myo Tun, "Sustainability of 5G green network toward D2D communication with RF-energy techniques," 2021 International Conference on Intelligent Technologies (CONIT), 2021, pp. 1–10, doi:10.1109/CONIT51480.2021.9498298

[10] Y. Wang, M. Narasimha, and R. W. Heath, "Mmwave beam prediction with situational awareness: A machine learning approach," in 2018 IEEE 19th International Workshop on Signal Processing Advances in Wireless Communications (SPAWC). IEEE, 2018, pp. 1–5.

[11] E. Balevi and J. G. Andrews, "Deep learning-based channel estimation for high-dimensional signals," *arXiv preprint arXiv:1904.09346*, 2019.

[12] H. He, C.-K. Wen, S. Jin, and G. Y. Li, "Deep learning-based channel estimation for beamspace mmwave massive mimo systems," *IEEE Wireless Communications Letters*, vol. 7, no. 5, pp. 852–855, 2018.

[13] M. S. Safari and V. Pourahmadi, "Deep ul2dl: Channel knowledge transfer from uplink to downlink," *arXiv preprint arXiv:1812.07518*, 2018.

[14] C.-K. Wen, W.-T. Shih, and S. Jin, "Deep learning for massive mimo csi feedback," *IEEE Wireless Communications Letters*, vol. 7, no. 5, pp. 748–751, 2018.

[15] M. Khani, M. Alizadeh, J. Hoydis, and P. Fleming, "Adaptive neural signal detection for massive mimo," *arXiv preprint arXiv:1906.04610*, 2019.

[16] G. Gao, C. Dong, and K. Niu, "Sparsely connected neural network for massive MIMO detection," EasyChair, Tech. Rep., 2018.

[17] M. Alrabeiah and A. Alkhateeb, "Deep learning for TDD and FDD massive MIMO: Mapping channels in space and frequency," *arXiv preprint arXiv:1905.03761*, 2019.

[18] C. Zhang, P. Patras, and H. Haddadi, "Deep learning in mobile and wireless networking: A survey," *IEEE Communications Surveys Tutorials*, vol. 21, pp. 2224–2287, 2019.

[19] L. Huang, A. D. Joseph, B. Nelson, B. I. Rubinstein, and J. D. Tygar, "Adversarial machine learning," in Proceedings of the 4th ACM workshop on Security and artificial intelligence, pp. 43–58, ACM, 2011.

[20] Devasis Pradhan, P.K. Sahu, K. Rajeswari, and Hla Myo Tun, "A study of localization in 5G green network (5G-GN) for futuristic cellular communication," The 3rd International Conference on Communication, Devices and Computing (ICCDC 2021), India, 16–18 August 2021.

[21] G. A. Plan, "An inefficient truth." Global Action Plan Report, 2007.

[22] Hla Myo Tun, "Radio network planning and optimization for 5G telecommunication system based on physical constraints", *Journal of Computer Science Research*, 3(1), January 2021. https://doi.org/10.30564/jcsr.v3i1.2701

[23] W. Song, F. Zeng, J. Hu, Z. Wang, and X. Mao, "An unsupervised learning-based method for multi-hop wireless broadcast relay selection in urban vehicular networks," *IEEE Vehicular Technology*, 865 Conference, vol. 2017.

[24] Pradhan, D., Sahu, P. K., Dash, A., & Tun, H. M. (2021, June). Sustainability of 5G Green Network toward D2D Communication with RF-Energy Techniques. In *2021 International Conference on Intelligent Technologies (CONIT)* (pp. 1–10). IEEE.

[25] M. S. Parwez, D. B. Rawat, and M. Garuba, "Big data analytics for user-activity analysis and user-anomaly detection in mobile wireless network," *IEEE Transactions on Industrial Informatics*, vol. 13, no. 4, pp. 2058–2065, 2017.

[26] L.-C. Wang and S. H. Cheng, "Data-driven resource management for ultra-dense small cells: An affinity propagation clustering approach," *IEEE Transactions on Network Science and Engineering*, vol. 4697, no. c, pp. 1–1, 2018.

[27] U. Challita, L. Dong, and W. Saad, "Deep learning for proactive resource allocation in LTE-U networks," in European Wireless 2017, 23rd European Wireless Conference, 2017.

[28] R. Pascanu, T. Mikolov, and Y. Bengio, "On the difficulty of training recurrent neural networks," Tech. Rep., 2013.

[29] Q.V. Le, N. Jaitly, and G.E. Hinton Google, "A simple way to initialize recurrent networks of rectified linear units," Tech. Rep., 2015.

[30] G. Alnwaimi, S. Vahid, and K. Moessner, "Dynamic heterogeneous learning games for opportunistic access in LTE-based macro/femtocell deployments," *IEEE Transactions on Wireless Communications*, vol. 14, no. 4, pp. 2294–2308, 2015.

[31] D. D. Nguyen, H. X. Nguyen, and L. B. White, "Reinforcement learning with network-assisted feedback for heterogeneous RAT selection," *IEEE Transactions on Wireless Communications*, vol. 16, no. 9, pp. 6062–6076, 2017.

[32] H. Sun, X. Chen, Q. Shi, M. Hong, X. Fu, and N. D. Sidiropoulos, "Learning to optimize: Training deep neural networks for interference management," *IEEE Transactions on Signal Processing*, vol. 66, no. 20, pp. 5438–5453, 2018.

[33] Y. Junhong and Y. J. Zhang, "Drag: Deep reinforcement learning based base station activation in heterogeneous networks," *IEEE Transactions on Mobile Computing*, vol. 19, no. 9, pp. 2076–2087, 2019.

[34] J. Liu, B. Krishnamachari, S. Zhou, and Z. Niu, "Deepnap: Data-driven base station sleeping operations through deep reinforcement learning," *IEEE Internet of Things Journal*, vol. 5, no. 6, pp. 4273–4282, 2018.

[35] M. Kozlowski, R. McConville, R. Santos-Rodriguez, and R. Piechocki, "Energy efficiency in reinforcement learning for wireless sensor networks," *arXiv preprint arXiv:1812.02538*, 2018.

[36] B. Matthiesen, A. Zappone, E. A. Jorswieck, and M. Debbah, "Deep learning for optimal energy-efficient power control in wireless interference networks," *arXiv preprint arXiv:1812.06920*, 2018.

[37] C. Gutterman, E. Grinshpun, S. Sharma, and G. Zussman, "Ran resource usage prediction for a 5G slice broker," in Proceedings of the Twentieth ACM International Symposium on Mobile Ad Hoc Networking and Computing. ACM, 2019, pp. 231–240.

[38] M. Chen, W. Saad, C. Yin, and M. Debbah, "Echo state networks for proactive caching in cloud-based radio access networks with mobile users," *IEEE Transactions on Wireless Communications*, vol. 16, no. 6, pp. 3520–3535, 2017.

[39] X. Lu, L. Xiao, C. Dai, and H. Dai, "Uav-aided cellular communications with deep reinforcement learning against jamming," *arXiv preprint arXiv:1805.06628*, 2018.

[40] M. Sadeghi and E. G. Larsson, "Physical adversarial attacks against end-to-end autoencoder communication systems," *IEEE Communications Letters*, vol. 23, no. 5, pp. 847–850, 2019.

15

Electrical Price Prediction using Machine Learning Algorithms

Swastik Mishra, Kanika Prasad, * **and Anand Mukut Tigga**

Department of Production and Industrial Engineering, National Institute of Technology, Jamshedpur, India

*kprasad.prod@nitjsr.ac.in

CONTENTS

DOI: 10.1201/9781003104858-15

255

15.1 Introduction

Energy is essential for the economic development of any country. In the case of developing countries, the energy sector assumes critical importance given the ever-increasing energy needs requiring huge investments to meet them. Electricity has influenced our day-to-day life and in industries, electricity provides power to large machines, which produce essential utilities and industrial and consumer products. Electricity provides the basis for the economic development of any country because a consistent and reliable power source creates a variety of businesses and job opportunities. It provides access to online resources and information.

Electricity is a secondary form of energy, because it's generated from the conversion of energy produced from different sources such as renewable, for example, solar energy, hydro energy, onshore offshore energy, and non-renewable such as coal, lignite, natural gases, and fossil fuels. It cannot be stored in large quantities due to its properties. Sometimes, there is a disbalance in the generation and supply of electrical energy. This causes fluctuation in the electricity prices. Extreme price volatility has forced power generators to hedge not only volume but also price risk. Price forecasts from a few hours to a few months ahead have been of particular interest to power portfolio managers. A power generation company capable of predicting volatile wholesale prices with a reasonable level of accuracy can adjust its strategy and its own production or consumption schedule to reduce the risk or maximize profits in day-ahead trading. It helps management take the right decisions and manage resources accordingly and work on the shortage. Lack of sufficient power generation capacity, poor transmission and distribution infrastructure, high costs of supply to remote areas, or simply lack of affordability for electricity are among the biggest hurdles for extending grid-based electricity. Forecasting of power price plays an essential role in the electricity industry, as it provides the basis for making decisions in power system planning and operation.

Electricity price forecasting is a time-series forecasting problem. Therefore, in this research work, electricity prices are forecast by applying machine learning (ML) algorithms with an open-source data set. The classical forecasting models and the combination of recurring neural networks (RNN) and convolutional neural networks (CNN) are used for price prediction. Classical models such as moving average, autoregressive moving average (ARMA), autoregressive integrated moving average (ARIMA), and long short-term memory (LSTM) are used for electricity price prediction. Electricity prices are governed by different variables known as price determinants. Therefore, after the application of classical forecasting models and RNN, multivariate forecasting models are applied on the same dataset. Consequently, the results obtained from the models applied are compared based on mean absolute percentage error (MAPE) and mean absolute error (MAE) values for both univariate models and multivariate models. The graphs of the actual price and forecasting price are also plotted to visualize the results obtained.

15.2   Literature Review

Researchers have already applied ML algorithms for forecasting the change in prices. Rolnick et al. defined the climate change problem and encouraged the use of ML algorithms for forecasting the adverse effects of climate change [1]. Kof Nti et al. presented a review on the classification of electrical energy and categorized the forecasting methods into industrial and artificial intelligence (AI) methods [2]. Aishwarya et al. applied ML models such as Bayesian linear regression, boosted decision tree regression, decision forest regression, and statistical analysis method for predicting the number of customers with internal data and external data in a ubiquitous environment [3]. Mir et al.'s comparative review shows that the time-series modeling approach has been extensively used for long- and medium-term forecasting. AI-based techniques for short-term forecasts remain prevalent in the literature [4]. Johannesen et al. compared various regression tools based on the lowest MAPE value and concluded that random forest regressor provides better short-term load prediction (30 min) and K-nearest neighbor (KNN) offers relatively better long-term load prediction (24 h) [5]. Ahmad and Chen determined that weather change is responsible for the change in energy consumption pattern in the domestic, commercial, and industrial sectors and concluded that LSBoost performance is more modest than the LMSR and NARM for monthly, seasonally, and yearly ahead intervals [6]. Fattah et al. developed an ARIMA model for demand forecasting of a product using the Box–Jenkins time-series approach [7]. For understanding the nature of time series and the objective of analysis, Athiyarath et al. explained and applied the concepts of LSTM, ARIMA, and CNN [8]. Kaushik et al. evaluated different statistical, neural, and ensemble techniques in their ability to predict patients' weekly average expenditures on certain pain medications. Two statistical models, persistence (baseline) and ARIMA, multilayer perceptron model, LSTM model and an ensemble model were used to predict the expenditures on two different pain medications [9]. Ahmad et al. used artificial neural networks (ANN) with a nonlinear autoregressive exogenous multivariable inputs model, multivariate linear regression model, and adaptive boosting model to predict energy demand in a smart grid environment [10]. Allee et al. used a data-driven approach for demand prediction using survey and smart meter data from 1,378 Tanzanian mini-grid customers. Applied support vector machines were used to predict building energy consumption in tropical regions [11]. Dong et al. presented support vector machine (SVM), a new neural network algorithm, to forecast energy consumption in buildings [12]. Jamii, an ARIMA model, was applied to model electrical energy consumption annually from 1971 to 2020 [13]. Hasanah et al. compared the performance of two methods in electric load demand forecasting. Genetic algorithm–support vector machine (GA–SVM) and ARIMA methods were applied for the prediction of daily load in Malang city, Indonesia [14]. Aurna et al. used ARIMA and Holt Winters model for the energy consumption of Ohio and Kentucky states and predicted the consumption considering different periods (daily, weekly, monthly) [15]. Velasco et al. used ARIMA and ANN as forecasting models to predict day-ahead electric load. Electric load data preparation, model implementation, and forecasting evaluation were done to assess if the prediction of the models met the acceptable error tolerance [16].

It is evident from the literature review that most of the research works have applied univariate forecasting models for price prediction. Since electrical price prediction is a multivariate forecasting problem, in this work, different univariate and multivariate forecasting models are applied. The results obtained are compared using MAE and MAPE metrics to determine the accuracy of prediction.

15.3 Methodology

This section presents the various univariate and multivariate forecasting models that were applied in this work.

15.3.1 Univariate Forecasting Models

15.3.1.1 Autoregressive Model

An autoregressive (AR) model predicts the future behavior based on past behavior. AR models are also called conditional models, Markov models, or transition models. They are used for forecasting when there is some correlation between values in a time series and the values that precede and succeed them.

An AR model is represented by Equation 15.1, where the value of outcome variable (P) at some point t in time is like "regular" linear regression, directly related to the predictor variable (Q). The simple linear regression and AR models differ in this respect that P is dependent on Q and also previous values of P.

$$P_t = \delta + \varphi_1 P_{t-1} + \varphi_2 P_{t-2} + \ldots + \varphi_p P_{t-p} + A_t \qquad (15.1)$$

where

- $P_{t-1}, P_{t-2} \ldots P_{t-p}$ are the past series values (lags),
- A_t is white noise (i.e., randomness), and
- δ is defined by the following equation:

$$\delta = (1 - \textstyle\sum \varphi_i)\mu \ (1 \leq i \leq p) \tag{15.2}$$

15.3.1.2 Moving Average (MA)

MA is one of the most foundational models for time-series forecasting, denoted as MA(a) where a is the order of the model. This is one of the basic statistical models that is a building block of more complex models. The MA model states that the current value is linearly dependent on the current and past error terms. Again, the error terms are assumed to be mutually independent and normally distributed, just like white noise. Equation 15.3 represents the predicted value using MA technique:

$$P_t = \mu + \alpha_1 + \varphi_1 \, \alpha_{t-1} + \varphi_2 \, \alpha_{t-2+} \cdots + \varphi_a \, \alpha_{t-a} \tag{15.3}$$

where P_t = present value
μ = mean of the series
φ = The magnitude of the impact of past errors on the present value
α = the present error term, and past error terms.
a = order of the series

15.3.1.3 ARMA

An ARMA model is used to describe weakly stationary stochastic time series in terms of two polynomials. The first of these polynomials is for autoregression, and the second for the MA. In this model, the impact of past lags along with the residuals is considered for predicting the future values as explained in the following equation:

$$P_t = \beta_1{}^* P_{t-1} + \alpha_1{}^* \varepsilon_{t-1} + \beta_2{}^* P_{t-2} + \alpha_2{}^*\varepsilon_{t-2} + \beta_3{}^*P_{t-3}$$
$$+ \alpha_3{}^*\varepsilon_{t-3} + \cdots + \beta^*P_{t-k} + \alpha^*\varepsilon_{t-k} \tag{15.4}$$

where
β = the coefficients of the AR model
α = the coefficients of the MA model.

15.3.1.4 ARIMA

ARIMA is a class of models that explain a given time series data based on its own previous values, that is, its own lags and the lagged forecast errors, so that Equations 15.5–15.7 can be used to forecast future values. The ARIMA model is used when the mean of the time series is not constant, that is, the time series is not stationary. Therefore, the data is prepared by a degree of differencing in order to make it stationary, that is, to remove trend and seasonal structures that negatively affect the regression model:

$$Z_t = a_{t+1} - a_t \qquad (15.5)$$

$$Z_t = \phi_1 {}^* Z_{t-1} + \theta_{1,} \epsilon_{t-1} + \epsilon_t \qquad (15.6)$$

$$a_1 = \sum_{i=1}^{k=1} Zk - i + a_1 \qquad (15.7)$$

where
a_t = data points on the non-stationary time series
$\phi_1 {}^* Z_{t-1}$ = represents the autoregressive part
$\theta_{1,} \epsilon_{t-1}$ = represents the moving average part
ϵ_t = error occurred

15.3.1.5 LSTM

LSTMs are special kind of RNNs, precisely designed to avoid the long-term dependency in problems. They can remember information for longer period of time with ease and it is their intended property. They possess a sigmoid neural net layer and a pointwise multiplication operation. The sigmoid layer outputs numbers between 0 and 1, describing how much of each component should be let through. A value of 0 means "let nothing through," while a value of 1 means "let everything through!" The vanishing gradient effect is red by implementing three gates along with the hidden state. The three implemented gates are commonly referred to as input, output, and forget gates. An LSTM has these gates to protect and control the cell state. Efficient applications of LSTM networks can be found in research fields such as human trajectory prediction, traffic forecasting, speech recognition, and weather prediction.

15.3.2 Multivariate Forecasting Models

15.3.2.1 DNN

DNNs are enhanced versions of the conventional neural networks. They are becoming very popular in recent times due to their outstanding performance to learn not only the nonlinear input–output mapping but also the underlying structure of the input data vectors. These types of models can approximate the behavior of any function (universal approximation theorem). The DNN output can be explained by the following equation:

$$p_i^l = f\left(\sum\nolimits_{j=1}^{J} \left(w_{i,k} q_k + b_i \right) \right) \qquad (15.8)$$

where p is the output of neuron of the current layer, q is the output of neuron of the previous layer, w is the set of weights, b is bias, f is nonlinear activation, i is current neuron, l is current layer, and k is the earlier layer.

15.3.2.2   CNN

CNN is traditionally designed for image detection systems but can also be used as time-series forecasting model. The neurons in CNN can hold high dimensional data and so this model performs better than the other neural networks. CNN models have three prominent layers: (i) convolutional layer, (ii) pooling layer, and (iii) flatten layer. A flatten layer is used between the convolutional layers and the dense layer to reduce the feature maps to a single one-dimensional vector.

15.3.2.3   CNN-LSTM

This architecture is also known as long-term recurrent convolutional network (LRCN) model. The CNN-LSTM architecture uses CNN layers for feature extraction on input data combined with LSTMs to support sequence prediction. A CNN-LSTM can be defined by adding CNN layers at the front end followed by LSTM layers with a dense layer at the output. These models were developed for visual time-series prediction problems.

15.3.2.4   CNN-LSTM-DNN

In order to perform complex tasks, the neural networks need to be trained deeper. When deeper networks start to converge, the problem of degradation occurs. In degradation, with the increase in network depth, accuracy gets saturated and then decreases rapidly. Unexpectedly, such degradation is not caused by overfitting, and adding more layers to a suitably deep model leads to higher training error. Skip connections, which are also known as shortcut connections, are introduced to overcome this problem. This type of connection skips some of the layers in the neural network and feeds the output of one layer as the input to the next layers. In this CNN-LSTM model, output is skipped to common DNN.

15.4   Electrical Demand Forecasting

15.4.1   Dataset

The dataset used in the study contains four years of electrical consumption, generation, pricing, and weather data for Spain. Consumption and generation data were retrieved from ENTSOE, a public portal for transmission service operator (TSO) data. The dataset is unique because it contains hourly data for electrical prices and the respective forecasts by the TSO for consumption and pricing. The work focuses on predicting electric prices better than that already forecast by the present data. The metrics we are using for comparison is MAPE and MAE defined below:

$$MAPE = \frac{100\%}{n} \sum\nolimits_{t=1}^{n} \left| \frac{At - Ft}{At} \right| \tag{15.9}$$

$$MAE = \frac{\sum_{i=1}^{n} |At - ft|}{n} \tag{15.10}$$

where
  A_t = actual value
  F_t = forecast value
  n = total number of datapoints

15.4.2 Data Preprocessing

In order to apply ML algorithms, the raw dataset is first cleaned and preprocessed. After cleaning the data, that is eliminating all the null values and noises, the dataset is converted to a usable format. The next step is feature scaling, after which the data is split into testing and training dataset.

For univariate time-series forecasting, a single feature is extracted and the value for total actual price is predicted. Subsequently, a min–max scaler is employed to scale the dataset. The dataset is divided into two parts: 80% as a training dataset and 20% as a testing dataset.

For multivariate time-series forecasting, multiple features, that is, energy consumption, price, day of the week, and month of the year are extracted. Then the scaling of the features is done using min–max scaler.

15.4.3 Experiment

15.4.3.1 Univariate Time-Series Forecasting

After data preprocessing and feature selection, the selected forecasting models are successfully applied using Jupyter notebook. Jupyter notebook is an open-source workbench used to create and share documents having codes, projects, equations, and visualization. The forecasted and actual prices are plotted in the graphs shown in Figure 15.1. From the figure, it can be concluded that the forecasted price is closest to actual price for the LSTM model. MAPE and MAE values obtained for each model is shown in Table 15.1.

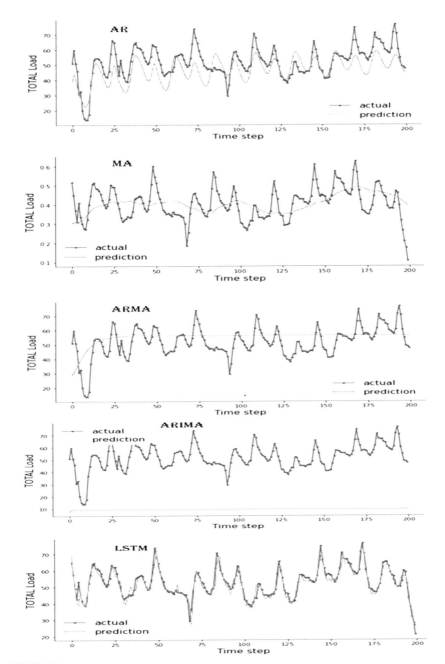

FIGURE 15.1
The actual and the forecasted price is plotted in the graph, predicted by the univariate forecasting models.

TABLE 15.1

The MAPE, RMSE, and MAE Values Obtained
for Each Univariate Model

Models	MAPE	MAE
Autoregressive	20.0674	12.4224
MA	11.7187	0.0484
ARMA	12.4534	20.4573
ARIMA	84.8715	55.6936
LSTM	4.1564	2.6708

15.4.3.2 Multivariate Time-Series Forecasting

In this section, all the multivariate forecasting models that are applied are explained. All the models have used the same final two DNN layers with dropout.

All the unique layers of all the models are stated below:

1. A three-layer DNN (one layer plus the common bottom two layers)
2. A CNN with two layers of 1D convolutions with max pooling
3. A LSTM with two LSTM layers
4. A CNN-stacked LSTM with layers from models 2 and 3 feeding into the common DNN layer. We are using the same layers from the CNN and LSTM model, stacking the CNN as input to the pair of LSTMs.
5. A CNN is stacked LSTM with a skip connection to the common DNN layer.

The same CNN and LSTM layers as the previous models are used this time with a skip connection direct to the common DNN layer.

Figure 15.2 shows that the LSTM appears to oscillate over a longer frequency compared to the other models. The CNN also seems to capture the intraday oscillations. CNN-stacked LSTM shows how these two attributes of the model's learning are combined. Table 15.2 shows the MAE values obtained by the different models.

15.5 Results and Discussions

15.5.1 For Univariate Models

After analyzing the results obtained after the application of the forecasting models, the LSTM forecasting model gives the most accurate results in

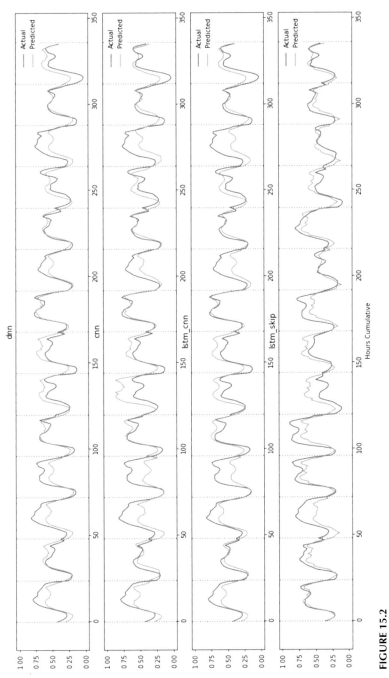

FIGURE 15.2

The prediction of price for two weeks using multivariate forecasting models.

TABLE 15.2

The MAPE Values Obtained by the Different Multivariate Models

MODELS	DNN	CNN	CNN-LSTM stacked	CNN-LSTM skip
MAPE	25.7879	21.4385	22.7875	23.6760
MAE	0.0877	0.0789	0.0828	0.0882

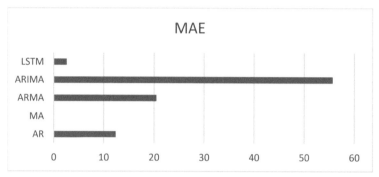

FIGURE 15.3
Comparison between the applied univariate models on the basis of MAE.

comparison to all the other forecasting models. If MAPE is taken as the metric of comparison, LSTM model gives the lowest MAPE. Lower MAPE shows that the LSTM forecasting results are closer to the actual prices as compared to other models. Figure 15.4 shows the MAPE values of all univariate models and it can be seen that MAPE of ARIMA is maximum, that is, ARIMA model's accuracy is the lowest. Figure 15.1 shows the comparison of actual prices and forecasted prices, and it can be seen that the prices forecasted by LSTM overlap with the actual price graph. Figure 15.3 shows the MAE obtained from all the univariate models applied and it can be seen MAE value for ARIMA is maximum.

15.5.2 For Multivariate Models

The CNN-stacked LSTM model gives the best results in comparison to LSTM skipped method. CNN and DNN also performed exceptionally well. The models are compared on the basis of MAPE and MAE metrics. Figure 15.5 shows the MAPE values of all models and it can be seen MAPE of CNN-stacked LSTM is lower as compared to LSTM skipped and MAPE of CNN is lowest. Figure 15.6 shows the MAE values obtained from the applied multivariate models. The MAE value obtained for CNN method is lowest among all the methods and therefore this model performed really well. The MAE

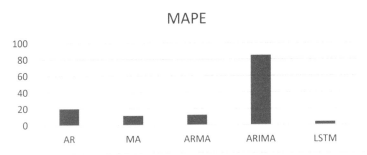

FIGURE 15.4
Comparison between the applied univariate models on the basis of MAPE.

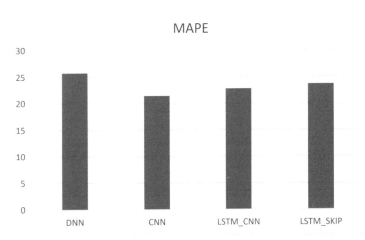

FIGURE 15.5
Comparison between the applied multivavariate models on the basis of MAPE.

of LSTM skipped is more than that of CNN stacked and that's why the fore-casted value obtained from CNN stacked is close to actual one in comparison to CNN skipped.

15.6 Conclusions

Electricity plays a key role in the economic development of any nation. All the power generation companies in the market aim to utilize all the available

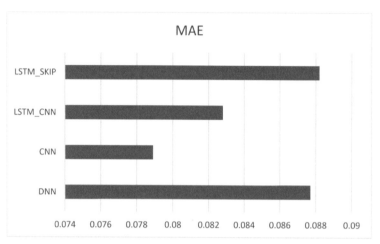

FIGURE 15.6
Comparison between the applied multivavariate models on the basis of MAE.

resources to the fullest, and improve the generation, consumption, and supply of electricity produced in order to maximize profit generation. Electricity price prediction plays a vital role in accomplishing the aims of the power companies. The accurate prediction of the electric prices enables the power managers to make decisions regarding the raw materials used and the production and supply of electricity. In this chapter, the idea is to apply different univariate and multivariate machine learning algorithms for electricity price prediction. Jupyter notebook, an open-source workbench, is used for coding and visualizing the results. Electric price predictions involve historic data of factors such as load and time in hours, days, and weeks. The machine learning algorithms perform well and handle a large amount of data. Out of the univariate models, LSTM outperforms all the other methods. Out of multivariate models, LSTM skip performs better than LSTM stacked. CNN and DNN give satisfactory results. As managers aim to understand the dynamics of the market, the methods discussed in this chapter can be used by them to create a concrete basis to make their decisions. Lower errors in forecasting lead to accurate decisions to be made in resource planning and capacity planning, which in turns improves the profit share of the power generation companies.

15.7 Limitations and Future Scope

In this research, the impact of weather on the forecasting model is not incorporated. Also, same techniques may be applied for electrical energy price forecasting in Indian context once the dataset is obtained.

References

[1] Rolnick, D., Donti, P.L., Kaack, L.H., Kochanski, K., Lacoste, A., Sankaran, K., Ross, A.S., Milojevic-Dupont, N., Jaques, N., Waldman-Brown, A., and Luccioni, A., 2019. Tackling climate change with machine learning. *arXiv preprint arXiv:1906.05433*.

[2] Nti, I.K., Teimeh, M., Nyarko-Boateng, O., and Adekoya, A.F., 2020. Electricity load forecasting: a systematic review. *Journal of Electrical Systems and Information Technology*, 7(1), 1–19.

[3] Aishwarya, K., Aishwarya Rao, Nikita Kumari, Akshit Mishra, and Rashmi, M.R., 2020. Food demand prediction using machine learning, *International Research Journal of Engineering and Technology*, 7(6), 3672–3675.

[4] Mir, A.A., Alghassab, M., Ullah, K., Khan, Z.A., Lu, Y., and Imran, M., 2020. A review of electricity demand forecasting in low and middle income countries: The demand determinants and horizons. *Sustainability*, 12(15), 5931.

[5] Johannesen, N.J., Kolhe, M., and Goodwin, M., 2019. Relative evaluation of regression tools for urban area electrical energy demand forecasting. *Journal of Cleaner Production*, 218, 555–564.

[6] Ahmad, T. and Chen, H., 2019. Nonlinear autoregressive and random forest approaches to forecasting electricity load for utility energy management systems. *Sustainable Cities and Society*, 45, 460–473.

[7] Fattah, J., Ezzine, L., Aman, Z., El Moussami, H., and Lachhab, A., 2018. Forecasting of demand using ARIMA model. *International Journal of Engineering Business Management*, 10, 1–9.

[8] Athiyarath, S., Paul, M., and Krishnaswamy, S., 2020. A comparative study and analysis of time series forecasting techniques. *SN Computer Science*, 1, 1–7.

[9] Kaushik, S., Choudhury, A., Sheron, P.K., Dasgupta, N., Natarajan, S., Pickett, L.A., and Dutt, V., 2020. AI in healthcare: time-series forecasting using statistical, neural, and ensemble architectures. *Frontiers in Big Data*, 3, 4.

[10] Ahmad, T. and Chen, H., 2018. Potential of three variant machine-learning models for forecasting district level medium-term and long-term energy demand in smart grid environment. *Energy*, 160, 1008–1020.

[11] Allee, A., Williams, N.J., Davis, A., and Jaramillo, P., 2021. Predicting initial electricity demand in off-grid Tanzanian communities using customer survey data and machine learning models. *Energy for Sustainable Development*, 62, 56–66.

[12] Dong, B., Cao, C., and Lee, S.E., 2005. Applying support vector machines to predict building energy consumption in tropical region. *Energy and Buildings*, 37(5), 545–553.

[13] Mohammed, J. and Mohamed, M., 2021. The forecasting of electrical energy consumption in morocco with an autoregressive integrated moving average approach, *Hindawi Mathematical Problems in Engineering*, 2021, Article ID 6623570, 9.

[14] Hasanah, R.N., Indratama, D., Suyono, H., Shidiq, M., and Abdel-Akher, M., 2020. Performance of genetic algorithm-support vector machine (GA-SVM) and autoregressive integrated moving average (ARIMA) in electric load forecasting. *Journal FORTEI-JEERI*, 1(1), 60–69.

[15] Aurna, Md. N.F., Rubel, T.M., Siddiqui, T.A., Karim, T., Saika, S., Md. Arifeen, M., Mahbub, T.N., Salim Reza, S.M., and Kabir H. 2021. Time series analysis of

electric energy consumption using autoregressive integrated moving average model and Holt Winters model, *TELKOMNIKA Telecommunication, Computing, Electronics and Control* 19(3), 991–1000.

[16] Velasco, L.C.P., Polestico, D.L.L., Macasieb, G.P.O., Reyes, M.B.V., and Vasquez Jr, F.B., 2018. Load forecasting using autoregressive integrated moving average and artificial neural network. *International Journal of Advanced Computer Science and Applications*, 9, 23–29.

16

Machine Learning Application to Predict the Degradation Rate of Biomedical Implants

Pradeep Bedi,[1] **Shyam Bihari Goyal,**[2] **Prasenjit Chatterjee,**[3] **and Jugnesh Kumar**[4]

[1]*Galgotais University, Greater Noida, Uttar Pradesh, India*

[2]*City University, Petaling Jaya, Malaysia*

[3]*MCKV Institute of Engineering, Liluah, Howrah West Bengal, India*

[4]*St. Andrews Institute of Technology & Management, Gurgaon, Haryana, India*

E-mail: 406130816@qq.com; drsbgoyal@gmail.com; dr.prasenjitchatterjee6@gmail.com; jugnesh@rediffmail.com

CONTENTS

16.1 Introduction

A bone fracture often termed as a broken bone, is a medical condition in which the shape or contour of bone changes due to impact to external forces or injuries under many biological as well as mechanical circumstances such as injuries during physical activities, vehicle accidents, accidental falls or due to weakening of bones because of aging as well an underlying disease [1]. Under fracture conditions, broken or cracked bone is stabilized and supported to handle the weight of the body for movement during the process of fracture healing. Some fractures are healed from outside of the body using plasters, but using them gives rise to critical issues. To reduce the risk of infection from external support, surgical procedures are performed to implant supports internally to stabilize fractured bones with some implants

DOI: 10.1201/9781003104858-16

TABLE 16.1

Different Types of Fractures and Their Internal Fixators

Types of Fracture	Internal Fixators
Skull fracture	Wires, pins, and plates
Craniofacial fracture	Wires, screws, and plates
Pelvic fracture	Screws, plates, and external fixators
Spinal fracture	Rods, pedicle screws, and plates
Upper and lowerlimb fracture	Plates and nails

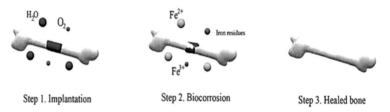

Step 1. Implantation   Step 2. Biocorrosion   Step 3. Healed bone

FIGURE 16.1
Bone healing procedure.

such as plates, screws, nails, or wires [2]. Different types of implant devices used for fracture healing are shown in Table 16.1.

These implants are of two types: non-biodegradable and biodegradable. Non-biodegradable implants have to be removed by second surgery after healing from fracture. But biodegradable implants degrade the inside body and therefore a second surgery is not needed after healing of the fractured bones. Degradation of biological implant materials occurs by an electro-chemical reaction in presence of an electrolyte, which results in the forma-tion of oxides, hydroxides, hydrogen gas, or other compounds, as shown in Figure 16.1. Metals are used as bone implant materials in a non-biodegradable implantation to fix cracked, deformed, worn-out, or broken bones.

The artificial replacements are made up of metals, polymers, or ceramic to be strong enough as well as flexible for everyday movement, as given in Figure 16.2. So, while choosing a material, it is necessary to ascertain its mechanical strength, flexibility, and biocompatibility. Biomaterials are syn-thetic materials, either degradable or non-degradable in nature, but they must possess good load-bearing capacity if they are to be used inside the human body [3–5].

Orthopedic biomaterials are implanted near the bone fracture to provide support and to heal the bone tissues. The end of fractured bones is connected to implant devices, which are fixed with metal pins or screws. After the healing of fractured bones, these screws and pins are removed. If these screws are made up of non-biodegradable materials, after healing, they need to be removed by a second operation.

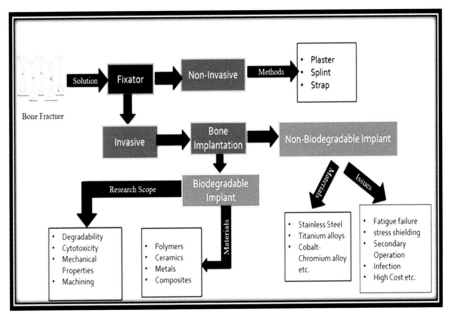

FIGURE 16.2
Bone fracture implant materials, issues, and research scope areas.

So, to avoid the second surgery, researchers are showing interest in developing biodegradable materials that can be used as implants. The bio-degradable screws, pins, nails, plates, and wires used as implants are generally left in the body after healing of bone fracture and they gradually degrade inside the body [6]. As the materials are left inside the body, the cytotoxic effect of biomedical materials have to be analyzed before they can be employed as implant devices. Researchers are exploring biodegradable materials that can improve the tissue regeneration process.

But firstly, it is necessary to understand what properties are essential for artificial replacement devices or implants. The required properties are as follows:

(i) The implants need to have a high biocompatibility.
(ii) The implants need to possess good mechanical properties, such as strength, during the process of bone healing.
(iii) The implants should show good thermal conductivity.
(iv) The implants should cause least friction at the implant location.
(v) The implants need to be resistant to mechanical shock.

Implants have to necessarily possess the above-mentioned properties during the replacement process [7]. While designing any bone fracture implant device, its surface and mechanical properties have to be studied. If the selection of the material or its alloy composition is not accurate, then it may lead to failure of the implant through loosening, osteolysis, wear, or toxic effect [8].

Subsequently, it is required to analyze the surrounding environment of implant device and tissues where the device is placed. Therefore, for better functioning of the biomedical implant, the choice of the appropriate material holds high importance.

The selection of materials for the implantation is a crucial step for successful long-term implants. The study of the implant material and its properties need to be focused upon for a successful treatment.

The characteristics of the implant materials studied are modulus of elasticity, compressive strength, tensile strength, shear strength, yield, and fatigue strength, ductility, hardness, corrosive properties (crevice corrosion, pitting corrosion, galvanic corrosion, electrochemical corrosion), surface tension, surface energy, and surface roughness [8]. Table 16.2 shows the material properties that can be considered for implantations.

The biodegradation progression is driven by three major factors such as chemical, mechanical, and biological interactions. In the case of the chemical-based deterioration of the polymer, the degradation rate is highly dependent on the polymer's composition, crystallinity, molecular structure as well as its hydrophobic and hydrophilic nature [2].

The chemical degradation process is achieved by breaking the polymer's molecular chains, breaking the cross-linking structure, or making interference

TABLE 16.2

Classification of Biomaterials

| Material Nature | Material Types | | |
	Metals	Ceramics	Polymers
Bio-toxic	Gold Co-Cr alloys Stainless steel Niobium Tantalum		Polyethylene Polyamide Polymethylmethacrylate Polytetrafluroethylene Polyurethane
Bio-inert	Commercially pure titanium Titanium alloy (Ti-6AL-4U)	Al oxide Zirconium oxide	
Bio-active		Hydroxyapatite tricalcium phosphate Bio-glass carbon silicon	
Biodegradable	Magnesium, Zinc, calcium, iron	Calcium phosphate, silica, alumina	Silk, collagen, polylactic acid

in its crystallinity. Bulk or surface degradation occurs in the body. In the case of bulk degradation, the rate of degradation is attained at a faster rate as in hydrophilic polymers by achieving their conversion into water-soluble materials. Surface degradation takes place in hydrophobic polymers in which it is intended to keep the inner structure intact and offers better control over degradation rates [3].

In the case of degradation at the biological level, the materials are exposed to the body fluids that result in changes in the chemical composition of the polymers. The degradation can through enzymatic, oxidation, or hydrolytic methods.

However, the interaction levels of the tissue and its behavior at the implant site depend on its physical, biological, and chemical nature. On the basis of their nature, implant device materials are categorized as:

Bio-toxic: Due to this nature, the surrounding tissues of implanted devices die.

Bioinert: Such materials show minimum interaction with surrounding cells or tissues. They don't show any adverse biological response.

Bioactive: Such materials are nontoxic in nature but show biological responses by releasing some chemical ions.

Bioresorbable/biodegradable: Such materials are nontoxic in nature and get dissolved inside the body. These materials replace or regenerate surrounding tissues biologically [5,6].

16.2 Related Work

Mahdi Dehestani et al. [9] experimentally investigated the mechanical properties and corrosion behavior of iron and hydroxyapatite (HA) composites for biodegradable implant applications. It was observed that the mechanical strength decreases with increasing HA content and decreasing HA particle size whereas their corrosion rates increased. Fe–2.5 wt% HA was finally created as the strongest composite.

Rakesh Rajan et al. [10] investigated the zinc–magnesium composite implant material for its mechanical, corrosion, and biological properties of magnesium. The mechanical strength is nearly that of the bone and the corrosion rate was observed to be 0.38 mm/year with a 12% elongation rate.

Richard et al. [11] investigated experimentally the corrosion properties of Fe–Mn–Si alloys for biodegradable medical implants. A corrosion rate of 0.24–0.44 mm/year was observed.

Tong et al. [12] studied the microstructure, mechanical properties, biocompatibility, and degradation behavior of Zn–Ge alloy for biodegradable implants. The experimental results show that it has a tensile strength of about 53.9 MPa and corrosion rate of 0.1272 mm/year, and elongation rate of about 1.2%.

Watrob et al. [13] experimentally designed a Zn-Ag-Zr alloy with enhanced strength as a potential biodegradable implant material that showed 17.1 ± 1.0 μm/year corrosion rate and also inhibits bacterial growth.

Dandan Xia et al. [14] investigated an alloy of Mg-Li-Ca for bone implant application, which was experimentally implemented on femurs of mice. This alloy shows excellent biocompatibility.

Suryavanshi et al. [15] developed a bone screw using a metal–polymer biodegradable composite for orthopedic applications. This was experimentally analyzed on rats and no toxic effects were observed.

Zhang et al. [16] analyzed the mechanical properties of magnesium nanocomposites, which were homogeneous exhibiting higher strength and ductility. These two properties reduced the elongation percentage of material.

Hongtao Yang et al. [17] prepared biodegradable zinc as biodegradable bone implants for load-bearing applications.

Razzaghi et al. [18] investigated the microstructure, mechanical properties, and biocompatibility of nanoalloy for implant applications. A major increase in the corrosion rate was observed.

Going by the literature, experimental analysis was undertaken for determining the suitable materials for implant designs and their properties. Some researchers are focused on the application of a machine learning approach for the optimization of performance parameters. Some contributions in the field of application of machine learning and implant design are discussed below.

Borgiani et al. [19] investigated and optimized the stress shielding effect of the hip prosthesis to achieve a better performance of implants. The optimization of the geometrical design of implants is achieved by machine learning techniques such as artificial neural network (ANN) and support vector machine (SVM). Parameters such as total stem length, thickness, and distance between the implant neck and stem surface are used for optimization. These parameters directly impact the stress shielding effect.

Chatterjee et al. [20] analyzed the strain deviation at different locations inside the body under two conditions, that is, before and after implantation. For the optimization of strain deviation, genetic algorithm was applied by finding the co-relation between the implant and their finite element simulation data by using artificial neural network (ANN). Different bone conditions were used to decide the specific guidelines while designing implant devices that must be patient-specific.

Niculescu et al. [21] studied the biomechanical behavior of orthopedic implants by applying deep learning and support vector machine. Further, mechanical testing was also performed for the analysis of implant stiffness.

Borjali et al. [22] and Bedi et al. [23] used machine learning methods to predict the wear rate of biomedical implants and to validate them by quantifying the prediction error.

16.3 Proposed Methodology

The following problems are faced:

Which material to select for different types of fractures.
Biocompatibility and mechanical properties of materials.
What composition materials are needed.
Take a quite long time to check degradation behavior.

Machine learning is a revolutionary application that can be applied in different ways in different industries. In this chapter, machine learning is applied for the prediction of the bone healing process after the occurrence of fractures in different parts of the body. For this, biodegradable material selection and its degradation rate are very important factors that need to be predicted and optimized for fast healing.

Some of the properties need to be studied while selecting the materials for manufacturing of implant devices.

Modulus of elasticity: Elasticity must be comparable to natural bone, that is, ~20GPa to for uniform stress distribution at implant for reduced relative movement at implant–bone interface.

Tensile strength: Implants should have high tensile strength, which lowers the stress at fracture interface.

Yield strength: Implants should have high yield strength to prevent brittle fracture under cyclic load.

Elongation rate: It is the measure for deformation in implants that determines ductility.

Degradation rate: Degradation rate should be related with the healing rate. Implant materials should be dependent on the length of time that it is necessary to keep them in the body.

Cytotoxicity: Implants should not be toxic to cells or tissues.

In this chapter, an algorithm is proposed for the prediction of material properties and its healing process according to the application area. Deciding the implant material and its composition is a very confusing task, such as which material is perfect to take as base material and its composition. The researcher

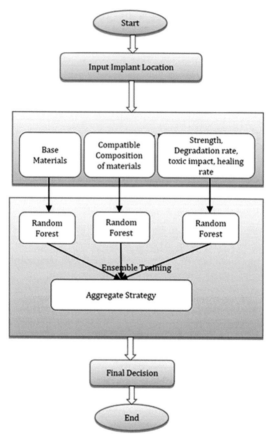

FIGURE 16.3
Proposed training process.

has to make samples and perform testing on them to know their mechanical and biological behavior. It is quite a long task to analyze the material. In any situation, if the material testing fails, the entire effort becomes worthless.

So, to reduce time consumption, the machine learning approach is used to predict the accuracy of material composition used for implant design. This process is performed in two steps: training and testing. Figure 16.3 illustrates the training process of the proposed model. Similarly, Figure 16.4 represents the testing process of the model.

Algorithm: Degradation of Behavior Prediction

> Initialization
> Select implant location
> Initialize the mechanical and biological properties

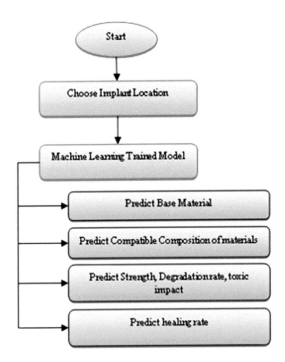

FIGURE 16.4
Testing process.

    Form S-groups
    Main Loop
    While iter < max iteration
    Select features
    Perform training process
    Aggregate training rules
    Update decision
    Final decision
    End

In ensemble machine learning training, each random forest module is trained independently with training samples and the data samples of each random forest are correctly classified. Similarly, all other data values are trained independently on individual random forest module and finally integrated as an ensemble or a combination of several random forest modules, which will expand the correctly classified area incrementally. This proposed ensemble random forest training module performs better in case of prediction problems. During the training phase of the proposed ensemble machine learning architecture, each random forest module is trained individually with training samples from the dataset. This makes each trained random forest module be different from each other. Each random forest module can be trained with different parameters of implant materials. Bagging, random selection, and boosting selection strategies can be used to select training samples

In this proposed architecture, bagging rules are taken as a base for ensemble module in which each module is trained individually and further they are aggregated by applying a combination method. During the testing phase, the aggregate strategy or voting strategy among all machine learning modules will decide the test data class label. In ensemble random forest architecture, *n* training sample sets are constructed with *n* individual modules. To achieve higher efficiency, different training sample sets are taken in order to improve the aggregation result with higher efficiency.

16.4   Conclusion

Biomedical implants are used in many applications such as hip replacement, femur bone replacement, and dental implants. Research has been ongoing on making better implant designs and selecting the best material composition for making implants last longer, in addition to reducing the body's reaction to their presence. For proper healing of the bone, implant devices are required to provide mechanical strength and stability. Choosing the proper material for making implant devices can improve osseointegration. Mechanical strength or tensile strength of the implants must be compatible with the natural bone. But selection of the material manually and its experimental analysis to determine its degradation rate take a very long time. The primary focus of this work is to use the power of machine learning for effective bone healing without any adverse effects. The use of machine learning will help in deciding the suitable implant material to enable fast healing of the bone fracture without any side effects.

References

[1]   Claes, L. and Ignatius, A. Development of new, biodegradable implants. *Chirurg.* 73, 990–996 (2002). https://doi.org/10.1007/S00104-002-0543-0.

[2]   Li, C., Guo, C., Fitzpatrick, V., Ibrahim, A., Zwierstra, M.J., Hanna, P., Lechtig, A., Nazarian, A., Lin, S.J., and Kaplan, D.L. Design of biodegradable, implantable devices towards clinical translation. *Nat. Rev. Mater.* 51(5), 61–81 (2019). https://doi.org/10.1038/s41578-019-0150-z.

[3]   Hofmann, G.O. Biodegradable implants in traumatology: A review on the state-of-the-art. *Arch. Orthop. Trauma Surg.* 1143 (114), 123–132 (1995). https://doi.org/10.1007/BF00443385.

[4] Liu, Y., Zheng, Y., and Hayes, B. Degradable, absorbable or resorbable—what is the best grammatical modifier for an implant that is eventually absorbed by the body? *Sci. China Mater.* 60, 377–391 (2017). https://doi.org/10.1007/S40 843-017-9023-9.

[5] Karpouzos, A., Diamantis, E., Farmaki, P., Savvanis, S., and Troupis, T. Nutritional aspects of bone health and fracture healing. *J. Osteoporos.* (2017). https://doi.org/10.1155/2017/4218472.

[6] Radha, R. and Sreekanth, D. Insight of magnesium alloys and composites for orthopedic implant applications—A review. *J. Magnes. Alloy.* 5, 286–312 (2017). https://doi.org/10.1016/J.JMA.2017.08.003.

[7] Wang, W., Han, J., Yang, X., Li, M., Wan, P., Tan, L., Zhang, Y., and Yang, K. Novel biocompatible magnesium alloys design with nutrient alloying elements Si, Ca and Sr: structure and properties characterization. *Mater. Sci. Eng. B.* 214, 26–36 (2016). https://doi.org/10.1016/J.MSEB.2016.08.005.

[8] Li, H., Yang, H., Zheng, Y., Zhou, F., Qiu, K., and Wang, X. Design and characterizations of novel biodegradable ternary Zn-based alloys with IIA nutrient alloying elements Mg, Ca and Sr. *Mater. Des.* 83, 95–102 (2015). https://doi.org/10.1016/J.MATDES.2015.05.089.

[9] Dehestani, M., Adolfsson, E., and Stanciu, L.A. Mechanical properties and corrosion behavior of powder metallurgy iron-hydroxyapatite composites for biodegradable implant applications. *Mater. Des.* 109, 556–569 (2016). https://doi.org/10.1016/J.MATDES.2016.07.092.

[10] Kottuparambil, R.R., Bontha, S., Rangarasaiah, R.M., Arya, S.B., Jana, A., Das, M., Balla, V.K., Amrithalingam, S., and Prabhu, T.R. Effect of zinc and rare-earth element addition on mechanical, corrosion, and biological properties of magnesium. *J. Mater. Res.* 33, 3466–3478 (2018). https://doi.org/10.1557/JMR.2018.311.

[11] Drevet, R., Zhukova, Y., Malikova, P., Dubinskiy, S., Korotitskiy, A., Pustov, Y., and Prokoshkin, S. Martensitic transformations and mechanical and corrosion properties of Fe-Mn-Si alloys for biodegradable medical implants. *MMTA.* 49, 1006–1013 (2018). https://doi.org/10.1007/S11661-017-4458-2.

[12] Tong, X., Zhang, D., Zhang, X., Su, Y., Shi, Z., Wang, K., Lin, J., Li, Y., Lin, J., and Wen, C. Microstructure, mechanical properties, biocompatibility, and in vitro corrosion and degradation behavior of a new Zn–5Ge alloy for biodegradable implant materials. *Acta Biomater.* 82, 197–204 (2018). https://doi.org/10.1016/J.ACTBIO.2018.10.015.

[13] Wątroba, M., Bednarczyk, W., Kawałko, J., Mech, K., Marciszko, M., Boelter, G., Banzhaf, M., and Bała, P. Design of novel Zn-Ag-Zr alloy with enhanced strength as a potential biodegradable implant material. *Mater. Des.* 183, 108154 (2019). https://doi.org/10.1016/J.MATDES.2019.108154.

[14] Xia, D., Liu, Y., Wang, S., Zeng, R.C., Liu, Y., Zheng, Y., and Zhou, Y. In vitro and in vivo investigation on biodegradable Mg-Li-Ca alloys for bone implant application. *Sci. China Mater.* 62, 256–272 (2018). https://doi.org/10.1007/S40 843-018-9293-8.

[15] Suryavanshi, A., Khanna, K., Sindhu, K.R., Bellare, J., and Srivastava, R. Development of bone screw using novel biodegradable composite orthopedic biomaterial: from material design to in vitro biomechanical and in vivo biocompatibility evaluation. *Biomed. Mater.* 14 (2019). https://doi.org/10.1088/1748-605X/AB16BE.

[16] Zhang, Z.Y., Guo, Y.H., Zhao, Y.T., Chen, G., Wu, J.L., and Liu, M.P. Effect of reinforcement spatial distribution on mechanical properties of MgO/ZK60 nanocomposites by powder metallurgy. *Mater. Charact.* 150, 229–235 (2019). https://doi.org/10.1016/J.MATCHAR.2019.02.024.

[17] Yang, H., Jia, B., Zhang, Z., Qu, X., Li, G., Lin, W., Zhu, D., Dai, K., and Zheng, Y. Alloying design of biodegradable zinc as promising bone implants for load-bearing applications. *Nat. Commun.* 11, 1–16 (2020). https://doi.org/10.1038/s41467-019-14153-7.

[18] Razzaghi, M., Kasiri-Asgarani, M., Bakhsheshi-Rad, H.R., and Ghayour, H. Microstructure, mechanical properties, and in-vitro biocompatibility of nano-NiTi reinforced Mg–3Zn-0.5Ag alloy: Prepared by mechanical alloying for implant applications. *Compos. Part B Eng.* 190, 107947 (2020). https://doi.org/10.1016/J.COMPOSITESB.2020.107947.

[19] Cilla, M., Borgiani, E., Martínez, J., Duda, G.N., and Checa, S. Machine learning techniques for the optimization of joint replacements: Application to a short-stem hip implant. *PLoS One.* 12 (2017). https://doi.org/10.1371/JOURNAL.PONE.0183755.

[20] Chatterjee, S., Dey, S., Majumder, S., RoyChowdhury, A., and Datta, S. Computational intelligence based design of implant for varying bone conditions. *Int. J. Numer. Method. Biomed. Eng.* 35 (2019). https://doi.org/10.1002/CNM.3191.

[21] Niculescu, B., Faur, C.I., Tataru, T., Diaconu, B.M., and Cruceru, M. Investigation of biomechanical characteristics of orthopedic implants for tibial plateau fractures by means of deep learning and support vector machine classification. *Appl. Sci.* 10, 4697 (2020). https://doi.org/10.3390/APP10144697.

[22] Borjali, A., Monson, K., and Raeymaekers, B. Predicting the polyethylene wear rate in pin-on-disc experiments in the context of prosthetic hip implants: Deriving a data-driven model using machine learning methods. *Tribol. Int.* 133, 101–110 (2019). https://doi.org/10.1016/J.TRIBOINT.2019.01.014.

[23] Bedi P., Goyal S.B., Rajawat A.S., Shaw R.N., and Ghosh A. (2022) A framework for personalizing atypical web search sessions with concept-based user profiles using selective machine learning techniques. In: Bianchini M., Piuri V., Das S., and Shaw R.N. (eds), *Advanced Computing and Intelligent Technologies. Lecture Notes in Networks and Systems*, vol. 218. Springer, Singapore. https://doi.org/10.1007/978-981-16-2164-2_23

17

Predicting the Outcomes of Myocardial Infarction Using Neural Decision Forest

Akashdeep Singh Chaudhary and Ravinder Saini

Department of Computer Science and Engineering, Chandigarh University
Mohali, Punjab, India
akashdeep1478@gmail.com; ravindersaini.cse@cumail.in

CONTENTS

DOI: 10.1201/9781003104858-17

17.1   Introduction

Myocardial infarction (MI) has become a serious silent killer in current times. It is very hard to predict the likely outcome for the patient suffering from MI. There are many factors involved such as diabetes, excessive alcohol consumption, high blood pressure, lack of exercise, high blood cholesterol, smoking, and lack of proper diet. Most of the patients suffering from MI have coronary artery disease (CAD). MI is usually caused due to the coronary artery blockage, which results from a breakage of an atherosclerotic plaque. So, an early stage-diagnosis with the help of tests such as electrocardiograms (ECGs), blood tests, and coronary angiography may help locate the presence of any disease. In general, chest pain is one of the symptoms that may occur due to MI and may travel to the shoulder, to the arm, and may even to the back. Feeling fainted, shortening of breath, cold sweat, or tiredness are a few other symptoms that can be observed in the patient's body. MI may lead to cardiac arrest, failure of the heart, or even an abnormal heartbeat.

Predicting MI outcomes is quite challenging because a many factors affect them. There is high mortality in the first year of patients suffering from acute myocardial infarction (AMI). A large number of people die from AMI even before reaching the hospitals. This is because of the huge uncertainty in the prediction of MI complications and outcomes. It can occur without complications or with complications. At the same time, it is observed that half the number of patients having acute or subacute MI have complications that result in worsening of the disease and even result in loss of life of the patient. It is hard even for experienced specialists to predict the complications. But with the help of a few techniques such as deep learning and using the previous data of the patients, it is quite feasible to some extent.

17.2   Neural Decision Forests

Neural decision forests (NDFs) are based upon two approaches. The first one is convolution neural networks (CNNs) and the second is random forests. To analyze visual imagery, the most commonly used artificial neural network is CNN. CNN works on the principle of feature maps, which are nothing but the translation equivariant responses produced by the filters that slide along input features. It corresponds to a shared-weight architecture of convolution kernels. CNNs are used in the field of computer vision, video processing, image processing, and natural language processing. The second approach consists of random forest, which is the method for classification-regression and many more, which is an ensemble learning technique. The random forest is the combination of interconnected trees, each of which is going to

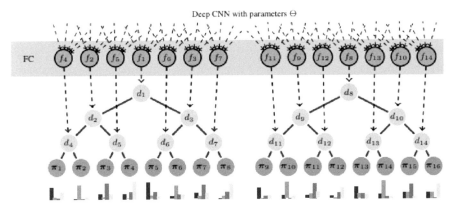

FIGURE 17.1
Architecture of the neural decision tree.

make predictions for the dataset provided as the input to it. Random forests or random decision forests create a large number of decision trees at the training time. In case of classification tasks, the class selected by most of the trees is the output of the random forest. Whereas in case of regression, the average prediction of the individual tree is the output. The purpose of using the concept of random forest in the NDF is to reduce overfitting to the training dataset. The deep neural decision forest (deep NDF) technique becomes the bridge between classification trees and the representation learning approach by training them in an end-to-end manner. Here, the concept of differentiable and the stochastic decision tree model comes into play. It propagates the representation learning organized in the initial layers of a (deep) CNN.

The deep NDF is the large number of interconnected decision trees, each of which makes a prediction for the given sample input and the average of all these predictions is made to get the random forest prediction. Figure 17.1 shows the architecture of a single neural decision tree. In other words, it can be said that random first uses a divide and conquer policy and has a simple model with very high performance.

In the NDF model, there are a number of neural decision trees that are trained at the same time. The average output of the trees is calculated as the final output of the NDF.

NDF consists of many neural decision trees in which each tree has to learn two types of weights, "pi" and "decision_fn." 'Pi' represents the probability distribution of the classes that are present in the tree leaves, whereas "decision_fn" represents the probability of going to each leaf node. There are four steps in the working of the neural decision tree. In the first step, the model takes the input features in the form of a single vector containing all the features of an instance in the batch. This vector is generated with the help of CNN. In the second step, the model randomly selects a subset of input features using "used_features_mask." In the third step, for each input instance taken from the second step, the model computes the probabilities (mu) to reach the tree leaves. The model iteratively performs a stochastic routing throughout the tree levels. Finally, in the fourth step, we get the final output by combining the class probabilities at the leaves with the probabilities of reaching the leaves.

17.3   Literature Review

Ibrahim el. al. [1] states that there is a need for accurate detection of AMI at an early stage so that the patient can get timely provision of medical intervention and is crucial for reducing the mortality rate. It is already known that machine learning has proved its potential in aiding the diagnosis of diseases. Lujain Ibrahim et al. [1] has used 713,447 extracted ECG samples along with the related auxiliary data from the longitudinal and comprehensive ECG-ViEW II database for predicting the AMI. The author has also conducted research using the XGBoost, which is a type of decision-tree-based model. The research revolves around creating a framework that can be used to detect AMI at an early stage. For the best-performing CNN, recurrent neural network (RNN) and XGBoost models, the prediction accuracy comes out to be 89.9, 84.6, and 97.5 percent, respectively, and ROC as 90.7, 82.9, and 96.5 percent (the curve areas) are achieved, respectively. The importance and use of the machine learning techniques are clearly proven to be of great value in the prediction of cardiovascular disease in the paper.

Lenselink et al. [2] studied the risk prediction models (RPMs) for CAD with the help of variables related to CAD. But the use of predictors in the clinical practice is very limited because of unavailability of a proper description of the model, method for external validation, and the head-to-head comparison. The author uses Tufts PACE CPM Registry and a systematic PubMed search to identify the RPMs for CAD prediction and all the selected models are externally verified in three large cohorts, namely, UK Biobank, LifeLines, and PREVEND. The author takes two endpoints, which are MI as a primary endpoint and CAD as a secondary endpoint, into consideration for validating every RPM externally. It consists of MI, coronary artery bypass grafting, and percutaneous coronary intervention. He calculates C-index (model discrimination), intercept and regression slope (calibration), and accuracy (Brier Score) to compare the selected RPMs. To estimate the calibration ability of an RPM, he used the method of linear regression analysis. In the paper, 28 RPMs were selected, but according to this paper, no best-performing RPM is identified as C-index of most of the RPMs is 0.706 ± 0.049, 0.778 ± 0.097, and 0.729 ± 0.074 for the prediction of MI in different cohorts of UK Biobank, LifeLines, and PREVEND, respectively. Therefore, research concludes that no particular RPM could be suggested for the prediction of risk of CADs.

Smith et al. [3] discusses various RPMs and carried out extensive research on AMI. He compares the published literature related to RPM for the prediction of AMI through March 2017 for 30-day hospital readmission among adults. He carried out research using 11 different studies on 18 unique RPMs having varied settings basically in the United States. Sixteen models out of 18 unique RPMs are specific for AMI. Out of 18 RPMs, six models chosen are based upon administrative data, four models are based upon electronic health record data, three models are based upon clinical hospital data, and the last five models are based upon the cardiac registry data. The range of the median overall observed is 10.6–21.0 percent (average 16.3 percent) across studies conducted. The most frequently used predictors in the models are demographics, comorbidities, and utilization metrics ranging from 7 to 37 in numbers. It is observed that the models have median C statistics as 0.65, having range 0.53–0.79. It is concluded from the studies that the models for the prediction of AMI have limited predictive ability and do not have certain generalizability because of the methodological limitations. Also, it can be said that no model provides the information related to the identification of AMI, based on which suitable action can be taken at an early stage and risk-stratification of patients suffering from AMI can be done before discharging from the hospital.

Mechanic et al. [4] discusses in detail about the acute MI and its results such as irreparable damage to the heart muscles because of an insufficient supply of oxygen. The patient may become prone to arrhythmias due to impairment in systolic function and diastolic because of MI. There can be a number of serious complications because of MI. Information related to many factors such as smoking, dyslipidemia, obesity, hypertension, diabetes mellitus, poor oral hygiene, and sedentary lifestyle. which affect MI is given in the paper. The paper discusses other causes of MI also which included trauma, vasculitis, drug use (cocaine), coronary artery anomalies, coronary artery emboli, aortic dissection, and excess demand on the heart (hyperthyroidism, anemia). According to this research, at least one-third of the patients die before reaching the hospital and another 40–50 percent of patients have a fatal outcome upon reaching the hospital. Another 5–10% of the patients would face the same outcome within the first year after having MI. The author makes the conclusion in the research that the physical examination of the patient must include the information related to vital signs, appearance of the patient, diaphoresis, and lung findings along with cardiac auscultation because it plays an important role to reveal more information. Ventricular arrhythmia, atrial fibrillation, or tachycardia may be revealed by observing the heart rate. The physical examination of the patient can reveal various other crucial points, such as observing blood pressure of patients can reveal hypotension, tachypnea and fever can be common, and because of the distended neck veins, there can be the right ventricular failure. Last, but not the least, if the patient has developed pulmonary edema, then he may have wheezing and rales.

Panju et al. [5] focuses on the features that help in increasing or decreasing the probability of AMI. The selected features are history, physical examination of the patient, and ECG data. The ECG data is included for research as the doctor usually interprets the results as the immediate initial clinical assessment of the condition of the patient. In the first step, three diagnostic groupings of the patients were made on the basis of acute chest pain and then compared the contrast with the categorization of chest pain whether the MI is present or not. The symptoms of MI are described briefly along with the signs of MI mechanism of chest pain and the conditions that are usually present with other symptoms related to MI. The paper discusses, in detail, the role of the accuracy and precision of history, the physical examination of the patient, and ECG data in the identification of MI. The clinical data related to these mentioned features along with their associated likelihood ratios (LRs) is considered for the prediction rules for AMI and are taken with a broad set of inclusion criteria. The paper provides the conclusion that the most crucial clinical feature is chest pain radiating toward the arm that increases the probability of MI. Precisely, patients having MI have two times the probability of having the chest pain radiating toward the left arm than the patients who are not having MI. Meanwhile, the probability of chest pain radiating toward the right arms is three times and the probability of chest pain radiating toward both the arms is seven times higher than the patients without MI.

Rossiev et al. [6] discusses the computer expert system to forecast four different types of complications that may appear to the patient suffering from the MI in the hospital period. The neural network used in the paper gathers the experience while training on the input data of real clinical cases. The dataset chosen contains four types of attributes that have the high priority or are more crucial. They divide the main task into eight subtasks, which contain four binary subtasks with the outcomes as 1 or 0 for having the complication or not having the complications, respectively, and four subtasks have the numerical output. For network training, backpropagation technique is used. The objective of training is to minimize the estimation function (global minimum taken). The conclusion made in this paper is about the great possibilities of creating a neural network expert system for predicting the complications of MI with the use of neural networks and accelerating the process of creating an expert system that does not require the mathematical algorithms for solving each task. The output from this expert system is probabilistic. The doctor can take help from the expert system by feeding the input to the expert system and make his or her own decision afterward.

17.4　Research Gap and Objective of the Research

From the literature review, it is found that less work has been done by taking the myocardial infarction dataset, which contains the 111 number of input parameters and with 12 outcome attributes (complications). It is very hard to establish the relationship between so many attributes and then make the prediction for the outcomes. The methods, techniques, approaches, and algorithms discussed in the literature review do not deal with the missing values in the dataset and thus give low accuracy and performance when dealing with the missing values. Also, the algorithms generally take longer to train the model, which is not good.

The main objective of this research is to make a model that solves the above three problems. The model should be able to work on a large number of input attributes as well as the model should not perform low on the dataset that contains missing values. The model should take less time to train on the dataset and with the least number of epochs, the model should give high performance.

17.5　Data Collection

MI complications dataset are taken from UCI Machine Learning [7]. The dataset has 1,700 instances with 124 unique attributes, but it has a few missing values also (about 7.6 percent). Figure 17.2 shows the first 20 instances of the dataset with the first header as the name of the attributes each starting with a unique ID.

Out of 124 attributes, the first column is specified for the patient's ID, columns 2–112 attributes are used as input data for prediction, and columns 113–124 are attributes to be predicted that are the possible outcomes (complications). The prediction is done for the time at the end of 72 h (72 h after admission to the hospital) and columns (2–112) are used as the input for prediction.

17.6　Data Analysis

As there are a lot of attributes used in the dataset, we have discussed only the attributes that are to be predicted (complications). We have given the details regarding the values of the attributes, their significance in relation to the MI,

```
1   ID,AGE,SEX,INF_ANAM,STENOK_AN,FK_STENOK,IBS_POST,IBS_NASL,GB,9
2   1,77,1,2,1,1,2,?,3,0,7,0,0,0,0,0,0,0,0,0,0,0,0,0,0,0,0,0,0,0,0,0,0,0,0,0,0,0,0,0,0,0,0,0,0,0
3   2,55,1,1,0,0,0,0,0,0,0,0,0,0,0,0,0,0,0,0,0,0,0,0,0,0,0,0,0,0,0,0,0,0,0,0,0,0,0,0,0,0,0,0,0,0,0
4   3,52,1,0,0,0,2,?,2,0,2,0,0,0,0,0,0,0,0,0,0,0,0,0,0,0,0,0,0,0,0,0,0,0,0,0,0,0,0,0,0,0,0,0,0,0
5   4,68,0,0,0,0,2,?,2,0,3,1,0,0,0,0,0,0,0,0,0,0,0,0,0,0,0,0,0,0,0,0,0,0,0,0,0,0,0,0,0,1,0,0
6   5,60,1,0,0,0,2,?,3,0,7,0,0,0,0,0,0,0,0,0,0,0,0,0,0,0,0,0,0,0,0,0,0,0,0,0,0,0,0,0,0,0,0,0,0
7   6,64,1,0,1,2,1,?,0,0,0,0,0,0,0,0,0,0,0,0,0,0,0,0,0,0,0,0,0,0,0,0,0,0,0,0,0,0,0,0,0,0,0,0
8   7,70,1,1,1,2,1,?,2,0,7,1,0,0,0,0,0,0,0,0,0,0,0,0,0,0,0,0,0,0,0,0,0,0,0,0,0,0,0,0,1,0,0
9   8,65,1,0,1,1,2,?,2,0,7,0,0,0,0,0,0,0,0,0,0,0,0,0,0,0,0,0,0,0,0,0,0,0,0,0,0,0,0,0,0,0,0,0,0
10  9,60,1,0,0,0,2,?,2,0,6,0,0,0,0,0,0,0,0,0,0,0,0,0,0,0,0,0,0,0,0,0,0,0,0,0,0,0,0,0,0,0,0,0,0
11  10,77,0,2,0,0,0,?,3,0,6,1,0,0,0,0,0,0,0,0,0,0,0,0,0,0,0,0,0,0,0,0,0,0,1,0,0,0
12  11,71,1,0,0,0,0,?,0,0,0,0,0,0,0,0,0,0,0,0,0,0,0,0,0,0,0,0,0,0,0,0,0,0,0,0,0,1,0,0,0
13  12,50,0,0,0,0,0,?,2,0,3,0,0,0,0,0,0,0,0,0,0,0,0,0,0,0,0,0,0,0,0,0,0,0,0,0,0
14  13,60,1,1,0,0,2,?,2,0,2,0,0,0,0,0,0,0,0,0,0,0,0,0,0,0,0,0,0,0,0,0,0,0,0,0
15  14,57,1,0,0,0,2,0,2,0,6,0,0,0,0,0,0,0,0,0,0,0,0,0,0,0,0,0,0,0,0,0,0,0,0,0,0
16  15,55,1,1,1,2,2,?,2,0,1,0,0,0,0,0,0,0,0,0,0,0,0,0,0,0,0,0,0,0,0,0,0,1,0,0
17  16,57,1,3,0,0,0,?,2,0,6,0,0,0,0,0,0,0,0,0,0,0,0,0,0,0,0,0,0,0,0,0,0,0,0
18  17,52,1,0,6,2,1,?,2,0,7,0,0,0,0,0,0,0,0,0,0,0,0,0,0,0,0,0,0,0,0,0,0,0,0
19  18,52,0,1,1,2,2,?,2,0,7,1,0,0,0,0,0,0,0,0,0,0,0,0,0,0,0,0,0,0,0,0,0,0,0,0
20  19,54,1,0,1,3,2,?,2,0,2,0,0,0,0,0,0,0,0,0,0,0,0,0,0,0,0,0,0,0,0,0,0,0,0
21  20,71,0,0,1,4,2,?,0,0,0,0,0,0,0,0,0,0,0,0,0,0,0,0,0,0,0,0,0,0,0,0,0
```

FIGURE 17.2
First 20 instances of the dataset.

the number of instances, and their percentage present in the dataset. In the dataset, the missing values are not present in the attributes that are to be predicted (complications), but these are present in other remaining attributes that are taken as input to the model. A brief description of the 12 attributes (113–124) used in the dataset, which are to be predicted, is given below.

17.6.1 Attribute 113 as FIBR_PREDS

It stands for atrial fibrillation having the values "0" and "1." Here, "0" stands for "no" and constitutes for 1,530 instances (90.00%), whereas "1" stands for "yes" and constitutes for 170 instances (10.00%) with no missing values.

17.6.2 Attribute 114 as PREDS_TAH

It stands for supraventricular tachycardia having the values "0" and "1." Here, "0" stands for "no" and constitutes for 1,680 instances (98.82 percent), whereas "1" stands for "yes" and constitutes for 20 instances (1.18 percent) with no missing values.

17.6.3 Attribute 115 as JELUD_TAH

It stands for ventricular tachycardia having the values "0" and "1." Here, "0" stands for "no" and constitutes for 1,658 instances (97.53 percent), whereas

"1" stands for "yes" and constitutes for 42 instances (2.47 percent) with no missing values.

17.6.4   Attribute 116 as FIBR_JELUD

It stands for ventricular fibrillation having the values "0" and "1." Here, "0" stands for "no" and constitutes for 1,629 instances (95.82 percent), whereas "1" stands for "yes" and constitutes for 71 instances (4.18 percent) with no missing values.

17.6.5   Attribute- 117 as A_V_BLOK

It stands for the third-degree AV block having the values "0" and "1." Here, "0" stands for "no" and constitutes for 1,643 instances (96.65 percent), whereas "1" stands for "yes" and constitutes for 57 instances (3.35 percent) with no missing values.

17.6.6   Attribute 118 as OTEK_LANC

It stands for the third-degree pulmonary edema having the values "0" and "1." Here "0" stands for "no" and constitutes for 1,541 instances (90.65 percent), whereas "1" stands for "yes" and constitutes for 159 instances (9.35 percent) with no missing values.

17.6.7   Attribute 119 as RAZRIV

It stands for myocardial rupture having the values "0" and "1." Here, "0" stands for "no" and constitutes for 1,646 instances (96.82 percent), whereas "1" stands for "yes" and constitutes for 54 instances (3.18 percent) with no missing values.

17.6.8   Attribute 120 as DRESSLER

It stands for Dressler syndrome having the values "0" and "1." Here, "0" stands for "no" and constitutes for 1,625 instances (95.59 percent), whereas "1" stands for "yes" and constitutes for 75 instances (4.41 percent) with no missing values.

17.6.9   Attribute 121 as ZSN

It stands for chronic heart failure having the values "0" and "1." Here, "0" stands for "no" and constitutes for 1,306 instances (76.82 percent), whereas "1" stands for "yes" and constitutes for 394 instances (23.18 percent) with no missing values.

17.6.10 Attribute 122 as REC_IM

It stands for relapse of the MI having the values "0" and "1." Here, "0" stands for "no" and constitutes for 1,541 instances (90.65 percent), whereas "1" stands for "yes" and constitutes for 159 instances (9.35 percent) with no missing values.

17.6.11 Attribute 123 as P_IM_STEN

It stands for post-infarction angina having the values "0" and "1." Here, "0" stands for "no" and constitutes for 1,552 instances (91.29 percent), whereas "1" stands for "yes" and constitutes for 148 instances (8.71 percent) with no missing valu

17.6.12 Attribute 124 as LET_IS

It stands for lethal outcome having the values "0," "1," "2," "3," "4," "5", "6," and "7." This attribute has eight possible values. Here, "0" stands for "alive" and constitutes for 1,429 instances (84.06 percent), "1" stands for "cardiogenic shock" and constitutes for 110 instances (6.47 percent), "2" stands for "pulmonary edema" and constitutes for 18 instances (1.06 percent), "3" stands for "myocardial rupture" and constitutes for 54 instances (3.18 percent), "4" stands for "progress of congestive heart failure" and constitutes for 23 instances (1.35 percent), "5" stands for "thromboembolism" and constitutes for 12 instances (0.71 percent), "6" stands for "asystole" and constitutes for 27 instances (1.59 percent), "7" stands for "ventricular fibrillation" and constitutes for 27 instances (1.59 percent), with no missing values.

17.7 Model Training

For training the model, we have used 80–20 split proportion for training and testing, respectively. The attributes of columns 113–124 are taken individually with the dataset of columns 1–112 for the prediction separately. We calculated the performance of the model at different instances by varying different parameters such as the number of epochs, the number of trees, and depth of trees. The objective of research is to find the best NDF model, which gives the highest performance for predicting the attributes 113–124 (complications) by varying different parameters that are discussed below.

17.7.1 Parameters Used for Training the Model

batch_size: 100
num_epoch: 10

17.7.2 Parameters for Training a Neural Decision Tree Model

depth: 5
used_feature_rate: 1.0

17.7.3 Parameters for Training a Neural Decision Forest Model

num_trees: 15
depth: 5
used_feature_rate: 0.50

By changing the value of the "used_features_rate" variable, we can change the number of features to be used by each tree. For training a single neural decision tree, the depth of the tree used is 5 and all the input features are selected as the input to the model by defining used_feature_rate equal to 1. For training the NDF model, the number of the trees (num_trees) is equal to 15, and the depth of each tree is equal to 5. We have randomly selected 50 percent of the input features for input to every tree by defining used_feature_rate equal to 0.5 while training the neural decision forest. The parameters with these values give the best performance.

17.8 Results

We evaluate our model on the input data at various parameters. Basically, we checked for the loss and sparse categorical accuracy of the model and found the best suitable parameters for the given objective and the input data. The best performance of the model is achieved by taking batch size as 100, and the number of epochs as 10.

In the case of a single neural decision tree model, we use the depth as 5 and we take all features for input features (used_feature_rate = 1.0).

In the case of neural decision forest, we use the number of trees as 10 (num_trees =10) and the depth as 10, while taking only half the number of total input features (used_feature_rate = 0.5), which are selected randomly.

The performance of a single neural decision tree model where we use all input features columns 2–112 for the prediction of attributes 113–124 (stated above) is shown in Table 17.1. The number of epochs used is 10, and the depth of the tree used is 5.

The performance of the NDF that consists of many neural decision trees where we use all input features columns 2–112 for the prediction of attributes 113–124 (stated above) is shown in Table 17.2. The number of epochs used is 10, number of trees used is 5, and the depth of trees used is 5.

TABLE 17.1

Performance of a Single Neural Decision Tree Model

S.No.	Attributes for the prediction	Loss	Sparse categorical accuracy	Test accuracy
1	FIBR_PREDS	0.5234	0.8059	80.59%
2	PREDS_TAH	0.1115	0.9824	98.24%
3	JELUD_TAH	0.1659	0.9647	96.47%
4	FIBR_JELUD	0.2706	0.9235	92.35%
5	A_V_BLOK	0.239	0.9353	93.53%
6	OTEK_LANC	0.6142	0.6176	61.76%
7	RAZRIV	0.5007	0.8412	84.12%
8	DRESSLER	0.1405	0.9824	98.24%
9	ZSN	0.6586	0.6441	64.41%
10	REC_IM	0.4775	0.8029	80.00%
11	P_IM_STEN	0.338	0.9	90.00%
12	LET_IS	0.865	0.8324	83.24%

TABLE 17.2

Performance of the Neural Decision Forest Model

S.No.	Attributes for the prediction	Loss	Sparse categorical accuracy	Test accuracy (%)
1	FIBR_PREDS	0.5429	0.8324	83.24
2	PREDS_TAH	0.1073	0.9824	98.24
3	JELUD_TAH	0.1603	0.9647	96.47
4	FIBR_JELUD	0.2711	0.9235	92.35
5	A_V_BLOK	0.2396	0.9353	93.53
6	OTEK_LANC	0.4113	0.8559	85.59
7	RAZRIV	0.5033	0.8412	84.12
8	DRESSLER	0.1536	0.9824	98.24
9	ZSN	0.5476	0.7941	79.41
10	REC_IM	0.4634	0.8441	84.41
11	P_IM_STEN	0.336	0.9	90.00%
12	LET_IS	0.9169	0.7353	73.53%

17.9   Conclusions

The use of machine learning in the field of health care and medicine has shown very good results and opens up tremendous possibilities in this field for the benefit of the society. Machine learning also provides clues related to diseases, which are quite difficult to predict with mathematical calculations alone, by using sophisticated techniques such as NDFs. By the use of the NDF method, the prediction of the possible outcomes (complications) of a disease becomes easy, and it provides high accuracy despite having missing values in the input data. This gives the opportunity for patients to become aware of the symptoms of the disease in advance and prepare accordingly or get treatment as soon as possible.

In future work, the creation of more complex machine learning algorithms is required to increase the efficiency of the prediction models. Moreover, there is a need to create an algorithm that can work efficiently, not only for the people of a particular geographical area, but for people who belong to different communities, different places, and different geographical locations. To avoid overfitting, the dataset should include the instances of the patients having variety in terms of the demographics, lifestyle, age, and medical history.

References

[1] Ibrahim L,Mesinovic M, Yang K-W, Eid MA. Explainable prediction of acute myocardial infarction using machine learning and Shapley values, *IEEE Access*, 2020;8: 210410–210417, , doi: 10.1109/ACCESS.2020.3040166.

[2] Lenselink C,Ties D,Pleijhuis R, van der Harst P. Validation and comparison of 28 risk prediction models for coronary artery disease, *Eur. J. Prev. Cardiol.*, 2021. doi: 10.1093/eurjpc/zwab095.

[3] Smith LN et al., Acute myocardial infarction readmission risk prediction models, *Circ. Cardiovasc. Qual. Outcomes*, 2018; 11(1). doi: 10.1161/CIRCOUTCOMES.117.003885.

[4] Mechanic OJ, Gavin M, Grossman SA. Acute Myocardial Infarction. [Updated 2022 May 9]. In: StatPearls [Internet]. Treasure Island (FL): StatPearls Publishing; 2022 Jan. Available from: www.ncbi.nlm.nih.gov/books/NBK459269/.

[5] Panju AA, Is this patient having a myocardial infarction? *JAMA*, 1998;280(14):1256, doi: 10.1001/jama.280.14.1256.

[6] Rossiev DA, Golovenkin SE, Shulman VA, Matjushin GV. Neural networks for forecasting of myocardial infarction complications, in The Second International Symposium on Neuroinformatics and Neurocomputers, pp. 292–298, doi: 10.1109/ISNINC.1995.480871.

[7] https://archive.ics.uci.edu/ml/datasets/Myocardial+infarction+compli cations

[8] Saleh M, Ambrose JA. Understanding myocardial infarction. F1000Res. 2018 Sep 3;7:F1000 Faculty Rev-1378. doi: 10.12688/f1000research.15096.1. PMID: 30228871; PMCID: PMC6124376.

[9] Wu J, Qiu J, Xie E, Jiang W, Zhao R, Qiu J, Zafar MA, Huang Y, Yu AC. Predicting in-hospital rupture of type A aortic dissection using Random Forest, *J. Thorac Dis.*, 2019;11(11):4634–4646.

[10] Chandler AB, Chapman I, Erhardt LR, et al. Coronary thrombosis in myocardial infarction. Report of a workshop on the role of coronary thrombosis in the pathogenesis of acute myocardial infarction. *Am. J. Cardiol.* 1974;34(7):823–833. doi: 10.1016/0002-9149(74)90703-6

[11] Thygesen K, Alpert JS, Jaffe AS, et al.: Third universal definition of myocardial infarction. Circulation. 2012;126(16):2020–35. doi: 10.1161/CIR.0b013e31826e1058

[12] DeWood MA, Spores J, Notske R, et al. Prevalence of total coronary occlusion during the early hours of transmural myocardial infarction. *N Engl J Med.* 1980;303(16):897–902. doi: 10.1056/NEJM198010163031601

[13] Ambrose JA, Najafi A. Strategies for the prevention of coronary artery disease complications: Can we do better? *Am J Med.* 2018; pii: S0002–9343(18)30382-6. doi: 10.1016/j.amjmed.2018.04.006

[14] Bassand JP, Hamm CW, Ardissino D, et al. Guidelines for the diagnosis and treatment of non-ST-segment elevation acute coronary syndromes: The task force for the diagnosis and treatment of non-ST-segment elevation acute coronary syndromes of the European Society of Cardiology. *Eur Heart J* 2007;28:1598–660.

[15] Mandelzweig L, Battler A, Boyko V, et al. The second Euro Heart Survey on acute coronary syndromes: Characteristics, treatment, and outcome of patients with ACS in Europe and the Mediterranean Basin in 2004. *Eur Heart J* 2006;27:2285–2293.

[16] Sanchis-Gomar F, Perez-Quilis C, Leischik R, Lucia A. Epidemiology of coronary heart disease and acute coronary syndrome. *Ann Transl Med* 2016;4:256–256.

[17] Frohlich ED, Quinlan PJ. Coronary heart disease risk factors: public impact of initial and later-announced risks. *Ochsner J* 2014;14:532–537.

18

Image Classification Using Contrastive Learning

Abhyuday Trivedi, Anjali Hembrom, Arkajit Saha, Tahreem Fatima, Shreya Dey, Monideepa Roy, and Sujoy Datta

School of Computer Engineering, KIIT, Deemed University, Bhubaneswar, Odisha, India

monideepafcs@kiit.ac.in

CONTENTS

DOI: 10.1201/9781003104858-18

18.1 Introduction

The problem of learning visual representations effectively without human supervision can be done by contrastive learning. This work has been based on learning visual representations by using the SimCLR framework [1].

Visual representations are image representation vectors on which supervised linear or unsupervised classifiers can be trained for accurate image recognition and classification. These representations can be learned by training deep learning models like ResNets on labeled datasets like ImageNet. But labeling and annotating data is a time-consuming and elaborate process and can be avoided by using alternate learning techniques like self-supervised learning where the training data is automatically labeled by finding and utilizing correlations between various input features.

Contrastive learning of visual representations can be done by efficiently finding similar and dissimilar images. For understanding contrastive representation learning, we interpreted the key elements of the SimCLR framework which show that:

- A combination of various different data augmentation operations is important for the contrastive prediction tasks to yield effective representations and stronger data augmentation benefits unsupervised contrastive learning more than supervised learning.
- Nonlinear transformation, which has the ability to learn between the illustration and the contrastive loss, is introduced and considerably improves the learned representations' quality.
- Normalized embeddings and an adjusted temperature parameter positively affects representation learning with contrastive cross-entropy loss.
- Larger batch sizes and longer training benefits contrastive learning more compared to its supervised counterpart.

The rest of the chapter will describe the background of previous work done in the field of using self-supervised and contrastive learning to achieve state-of-the-art results; how SimCLR framework has been used and implemented in this work; visualization of the results obtained from the implementation and future scope and applications of our work.

18.2 Background

In this chapter, we explored and highlighted how a simple framework for contrastive learning of visual representations referred to as SimCLR [1] helps us to learn the visual representations. We have used ResNet-18 and trained our models for 20 epochs to get a particular convergence and show how SimCLR benefits more from unsupervised contrastive learning than supervised learning.

The idea of classifying representations of an image on the basis of similarity with each other under small transformations dates long back. Dating back to those approaches [2], which used to learn representations by contrasting positive pairs against negative pairs, SimCLR extended them by highlighting and implementing recent advances in data augmentation, neural network architecture, and contrastive loss. The individual components of SimCLR have previously been researched upon, but here the improved and integrated version, a result of a combination of different design choices, which provides better performance has been used and described in the Results and Discussion section of the chapter.

SimCLR has been used as the basis method of contrastive learning as some ways by which SimCLR is better for visual representation than previously proposed contrastive representation learning methods are:

- DIM/AMDIM [3][4] modified a ResNet and placed restrictions on the receiving fields of the convolutional neural network (e.g., 3 × 3 Convs were replaced with 1 × 1 Convs). In SimCLR framework, the decoupled task, which has to be predicted, and neural network base encoder architecture uses the final vector representations of two augmented image vectors for prediction, thus needing powerful ResNets.

- CPC v1 and v2 [5][6] used a deterministic strategy to define the context prediction task and a context aggregation network (for instance, a Pixel-CNN). SimCLR framework does not need such a context aggregation network as it decouples the prediction task and the neural network base encoder. In both these above pointers discussing different methods, it is important to note that SimCLR uses simplified data augmentation and that its NT- Xent Loss function uses normalization and adjustment of temperature parameter to give better similarity scores.

- InstDisc, MoCo, and PIRL [7][8][9] generalized the Exemplar approach originally proposed by [10] and leveraged an explicit memory bank. SimCLR, on the other hand, doesn't use a memory bank but in-batch negative example sampling is utilized.

- CMC [11] used a separate network for each view, whereas SimCLR only shares only a single network among all the random augmented views.

- Similarity between augmented and un-augmented copies of the same original image are enhanced to the maximum possible value, but SimCLR applies data augmentation fairly to both symmetric branches of its framework [12].

18.3　Implementation

18.3.1　Self-Supervised Learning

Self-supervised learning method is used to train computers to do tasks without the need to provide labeled data manually (Figure 18.1). It is a subset of un-supervised learning where results are derived by information labeling, categorizing, and finally analyzing it. The conclusions are drawn by the machine itself on the basis of connections and correlations. The system learns and attains the capability to know the different parts of any object by encoding it, so that recognition can be done from any angle. Only then can the object be classified correctly and provide context for analysis to come up with the desired output.

18.3.2　Contrastive Learning

Contrastive learning is a method used for finding similar and dissimilar things by training a ML model to learn classification between similar and dissimilar images (Figure 18.2). Contrastive learning contrasts positive pairs against negative pairs and learns representations. This method of learning

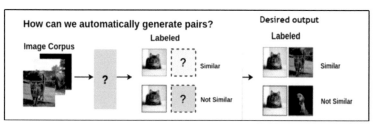

FIGURE 18.1
Self-supervised learning results.

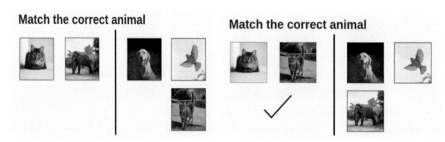

FIGURE 18.2
Expected contrastive learning prediction output.

first learns generic representations of images from an unlabeled dataset and fine-tuned with a small dataset of labeled images for classification purposes. Contrastive learning can be applied by

- generation of certain sized image batches,
- application of a transformation function to get a pair of two modified images,
- passing the pair obtained through an encoder to get image representations,
- passing the representations of the two augmented images through a nonlinear dense layer, and
- getting an embedding vector for each augmented image in the batch.

The agreement between two augmented versions of an image is determined using loss function's cosine similarity feature. To find the probability of how similar these two images are, a soft-max function is applied.

18.3.3 SimCLR

SimCLR maximizes similarity value between differently augmented versions of same image data by contrastive loss in latent space and learns visual representations (Figure 18.3). The framework includes the following major processes:

- To begin, SimCLR randomly draws examples from the original dataset, transforms each example twice using a combination of simple augmentations (random cropping, resize, and color distortion), creating

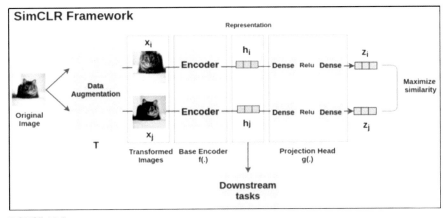

FIGURE 18.3
SimCLR framework.

two sets of corresponding views. This comprises the data augmentation module.

- SimCLR then computes the image representation vectors from the augmented images using a neural network base encoder function and for this purpose convolutional neural network (CNN) variant based on the ResNet-18 architecture is chosen.

- It maps a nonlinear projection of the image representation using a neural network projection head of a fully connected network (i.e., MLP), which amplifies the invariant features and maximizes the ability of identification of different transformations of the same image.

- Then, a stochastic gradient descent is used to update both CNN and MLP in order to minimize the loss function of the contrastive objective for contrastive prediction task.

These components of the framework have been implemented and highlighted in the following subsections.

18.3.3.1 Dataset

We have used a manually created ImageNet dataset containing 1250 images for training (250 images for each of the 5 categories) and 250 images for test (50 images for each of the five categories). The five categories used for SimCLR framework analysis are car, dog, bear, donut, and jean.

18.3.3.2 Data Augmentation

Data augmentation has previously been used for supervised or unsupervised representation learning [4,6,13] for contrastive prediction tasks by making changes in the architecture. In SimCLR application, generally, for a batch of N images, on applying the below mentioned composition of data augmentation operations, $2N$ augmented images are obtained as shown in Figure 18.4. Then, for a particular positive pair of images (i, j) from the $2N$ images, $2(N-1)$ images are considered as negative pair (i, j) examples as displayed in Figure 18.5.

In our analysis, as ImageNet images are always of different sizes; each of the images in our dataset went through the first data augmentation operation and were cropped and resized to size 224 × 224 (Figure 18.6). This was done by standard random cropping [14] of random size 1.0 in area, of original size and then resizing of the cropped image. Thus, we obtained the first set of augmented images. This implementation is carried out in Pytorch and makes use of RandomResizedCrop class of torch package transforms of Pytorch library as shown in the pseudocode in Figure 18.7. Additionally, in another data augmentation operation, color distortion was done to get another set

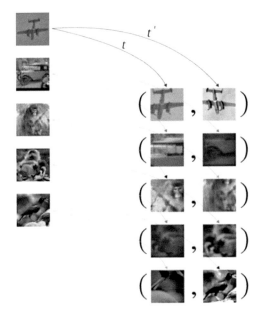

FIGURE 18.4
2N images formed from batch of N images on performing data augmentation operations.

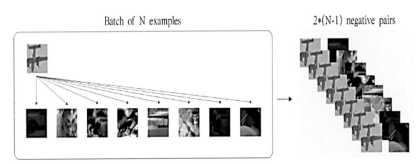

FIGURE 18.5
From the 2N images, 2(N–1) images considered as negative pairs.

of augmented images where color distortion is composed of strong color jittering controlled by using strength parameters and color dropping.

18.3.3.3 Extraction of Representation Vectors with Neural Network Encoder

The next component of the framework feeds each positive pair obtained from previous process in any neural network (standard ResNets) encoder

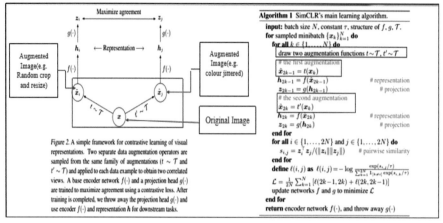

FIGURE 18.6
Data augmentation highlighted in main SimCLR algorithm.

```
# A function to perform color distortion in images
# It is used in SimCLR alongwith random resized cropping
# Here, s is the strength of color distortion.

def get_color_distortion(s=1.0):
    color_jitter = T.ColorJitter(0.8 * s, 0.8 * s, 0.8 * s, 0.2 * s)
    rnd_color_jitter = T.RandomApply([color_jitter], p=0.8)

    # p is the probability of grayscale, here 0.2
    rnd_gray = T.RandomGrayscale(p=0.2)
    color_distort = T.Compose([rnd_color_jitter, rnd_gray])

    return color_distort

# this is the dataset class

class MyDataset(Dataset):
    def __init__(self, root_dir, filenames, labels, mutation=False):
        self.root_dir = root_dir
        self.file_names = filenames
        self.labels = labels
        self.mutation = mutation

    def __len__(self):
        return len(self.file_names)

    def tensorify(self, img):
        res = T.ToTensor()(img)
        res = T.Normalize((0.5, 0.5, 0.5), (0.5, 0.5, 0.5))(res)
        return res
```

FIGURE 18.7
Data augmentation displayed in pseudo code.

SimCLR - Algorithm

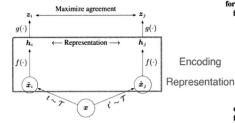

Algorithm 1 SimCLR's main learning algorithm.

input: batch size N, constant τ, structure of f, g, $\mathcal{T}$.
for sampled minibatch $\{x_k\}_{k=1}^N$ **do**
  for all $k \in \{1, \dots, N\}$ **do**
    draw two augmentation functions $t \sim \mathcal{T}, t' \sim \mathcal{T}$
    # the first augmentation
    $\tilde{x}_{2k-1} = t(x_k)$
    $h_{2k-1} = f(\tilde{x}_{2k-1})$ # representation
    $z_{2k-1} = g(h_{2k-1})$ # projection
    # the second augmentation
    $\tilde{x}_{2k} = t'(x_k)$
    $h_{2k} = f(\tilde{x}_{2k})$ # representation
    $z_{2k} = g(h_{2k})$ # projection
  end for
  for all $i \in \{1, \dots, 2N\}$ and $j \in \{1, \dots, 2N\}$ **do**
    $s_{i,j} = z_i^\top z_j/(\|z_i\|\|z_j\|)$ # pairwise similarity
  end for
  define $\ell(i,j)$ **as** $\ell(i,j) = -\log \frac{\exp(s_{i,j}/\tau)}{\sum_{k=1}^{2N} \mathbb{1}_{[k \neq i]} \exp(s_{i,k}/\tau)}$
  $\mathcal{L} = \frac{1}{2N} \sum_{k=1}^N [\ell(2k-1, 2k) + \ell(2k, 2k-1)]$
  update networks f and g to minimize $\mathcal{L}$
end for
return encoder network $f(\cdot)$, and throw away $g(\cdot)$

Encoding
Representation

Figure 2. A simple framework for contrastive learning of visual representations. Two separate data augmentation operators are sampled from the same family of augmentations ($t \sim \mathcal{T}$ and $t' \sim \mathcal{T}$) and applied to each data example to obtain two correlated views. A base encoder network $f(\cdot)$ and a projection head $g(\cdot)$ are trained to maximize agreement using a contrastive loss. After training is completed, we throw away the projection head $g(\cdot)$ and use encoder $f(\cdot)$ and representation h for downstream tasks.

FIGURE 18.8
Encoder function for representation vectors in main SimCLR algorithm.

```
# defining our deep learning architecture
resnet = resnet18(pretrained=False)

classifier = nn.Sequential(OrderedDict([
    ('fc1', nn.Linear(resnet.fc.in_features, 100)),
    ('added_relu1', nn.ReLU(inplace=True)),
    ('fc2', nn.Linear(100, 50)),
    ('added_relu2', nn.ReLU(inplace=True)),
    ('fc3', nn.Linear(50, 25))
]))

resnet.fc = classifier


# moving the resnet architecture to device
resnet.to(device)
```

FIGURE 18.9
ResNet-18 with top layer replaced by fully connected layers and last layer replaced with nonlinear classifier.

shown as f to obtain representation vectors represented as shown in Figure 18.8. The representation vectors are the output obtained after the average pooling layer.

We modified the SimCLR framework to compute the image representation vectors from the augmented images using a neural network base encoder function, but for this analysis, convolutional neural network (CNN) variant based on the ResNet-18 [15] architecture is chosen where the top layer is replaced by some fully connected layers as shown in Figure 18.9.

18.3.3.4 Nonlinear Projection Head

This component of the framework uses the representation vectors h obtained from the standard encoding layer f and uses a nonlinear projection head g (Figure 18.10). As the hidden layer before nonlinear projection head gives a better representation than the layer after, the hidden layer is preferred for representation. The vectors obtained as z are trained to remain unaltered after data transformation. On using the projection head g, more information is formed and retained at h; as the representation h is used before nonlinear projection due to loss of more information at g because of contrastive loss [1].

As already mentioned in the above component of the framework being analyzed, we use ResNet-18 as the base encoder network and a two-layer nonlinear MLP projection head is used to project the representation further. Figure 18.11 shows how the last layer and ReLU activation layer is removed incrementally and visualized using tSNE visualization while the Figure 18.12

SimCLR - Algorithm

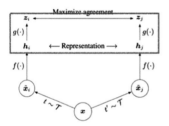

Figure 2. A simple framework for contrastive learning of visual representations. Two separate data augmentation operators are sampled from the same family of augmentations ($t \sim \mathcal{T}$ and $t' \sim \mathcal{T}$) and applied to each data example to obtain two correlated views. A base encoder network $f(\cdot)$ and a projection head $g(\cdot)$ are trained to maximize agreement using a contrastive loss. After training is completed, we throw away the projection head $g(\cdot)$ and use encoder $f(\cdot)$ and representation h for downstream tasks.

Algorithm 1 SimCLR's main learning algorithm.
___
input: batch size N, constant τ, structure of f, g, $\mathcal{T}$.
for sampled minibatch $\{x_k\}_{k=1}^{N}$ **do**
  for all $k \in \{1, \ldots, N\}$ **do**
    draw two augmentation functions $t \sim \mathcal{T}, t' \sim \mathcal{T}$
    # the first augmentation
    $\tilde{x}_{2k-1} = t(x_k)$
    $h_{2k-1} = f(\tilde{x}_{2k-1})$ # representation
    $z_{2k-1} = g(h_{2k-1})$ # projection
    # the second augmentation
    $\tilde{x}_{2k} = t'(x_k)$
    $h_{2k} = f(\tilde{x}_{2k})$ # representation
    $z_{2k} = g(h_{2k})$ # projection
  end for
  for all $i \in \{1, \ldots, 2N\}$ and $j \in \{1, \ldots, 2N\}$ **do**
    $s_{i,j} = z_i^{\top} z_j / (\|z_i\| \|z_j\|)$ # pairwise similarity
  end for
  define $\ell(i, j)$ as $\ell(i,j) = -\log \frac{\exp(s_{i,j}/\tau)}{\sum_{k=1}^{2N} \mathbb{1}_{[k \neq i]} \exp(s_{i,k}/\tau)}$
  $\mathcal{L} = \frac{1}{2N} \sum_{k=1}^{N} [\ell(2k-1, 2k) + \ell(2k, 2k-1)]$
  update networks f and g to minimize $\mathcal{L}$
end for
return encoder network $f(\cdot)$, and throw away $g(\cdot)$

Nonlinear

Projection

FIGURE 18.10
Nonlinear projection head used after representation in main SimCLR algorithm.

```
# Removing the last layer and the relu layer, we remove layers incrementally and look t-SNE visualizations
resnet.fc = nn.Sequential(*list(resnet.fc.children()))[:-2])

if TSNEVIS:
    for (_, sample_batched) in enumerate(dataloader_training_dataset):
        x = sample_batched['image']
        x = x.to(device)
        y = resnet(x)
        y_tsne = tsne.fit_transform(y.cpu().data)
        labels = sample_batched['label']
        plot_vecs_n_labels(y_tsne,labels,'tsne_train_second_last_layer.png')
```

FIGURE 18.11
Removal of last and ReLU layer and the method to see the tSNE visualization.

shows the removal of the last layer of the projection head and its affect can be visualized using tSNE visualization.

18.3.3.5 Normalized Temperature-Scaled Cross-Entropy Loss Function for Contrastive Prediction

Our motive is minimizing normalized temperature-scaled cross-entropy loss or NT-Xent Loss [1] for positive pair of examples, which maximizes the similarity between vectors z_i and z_j and at the same time makes the dissimilarity with all other vectors evident. Figure 18.13 shows the asymmetric loss function used here.

For maximizing the agreement between image vectors z_i and z_j for analysis, we used NT-Xent Loss optimized using SGD optimizer with learning rate 1.2 (=√Batch size * 0.075) and weight decay value10^{-6} (Figure 18.14). We

```
# removing one more layer, our entire projection head will be removed after this
resnet.fc = nn.Sequential(*list(resnet.fc.children())[:-1])

if TSNEVIS:
    for (_, sample_batched) in enumerate(dataloader_training_dataset):
        x = sample_batched['image']
        x = x.to(device)
        y = resnet(x)
        y_tsne = tsne.fit_transform(y.cpu().data)
        labels = sample_batched['label']
        plot_vecs_n_labels(y_tsne,labels,'tsne_hidden_train.png')
```

FIGURE 18.12
Removal of last layer of projection head and the method to see the tSNE visualization.

SimCLR - Algorithm

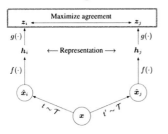

Figure 2. A simple framework for contrastive learning of visual representations. Two separate data augmentation operators are sampled from the same family of augmentations ($t \sim \mathcal{T}$ and $t' \sim \mathcal{T}$) and applied to each data example to obtain two correlated views. A base encoder network $f(\cdot)$ and a projection head $g(\cdot)$ are trained to maximize agreement using a contrastive loss. After training is completed, we throw away the projection head $g(\cdot)$ and use encoder $f(\cdot)$ and representation h for downstream tasks.

Algorithm 1 SimCLR's main learning algorithm.

input: batch size N, constant τ, structure of f, g, $\mathcal{T}$.
for sampled minibatch $\{x_k\}_{k=1}^N$ **do**
  for all $k \in \{1, \ldots, N\}$ **do**
    draw two augmentation functions $t \sim \mathcal{T}$, $t' \sim \mathcal{T}$
    # the first augmentation
    $\tilde{x}_{2k-1} = t(x_k)$
    $h_{2k-1} = f(\tilde{x}_{2k-1})$ # representation
    $z_{2k-1} = g(h_{2k-1})$ # projection
    # the second augmentation
    $\tilde{x}_{2k} = t'(x_k)$
    $h_{2k} = f(\tilde{x}_{2k})$ # representation
    $z_{2k} = g(h_{2k})$ # projection
  end for
  for all $i \in \{1, \ldots, 2N\}$ and $j \in \{1, \ldots, 2N\}$ **do**
    $s_{i,j} = z_i^\top z_j / (\|z_i\| \|z_j\|)$ # pairwise similarity
  end for
  define $\ell(i, j)$ as $\ell(i, j) = -\log \frac{\exp(s_{i,j}/\tau)}{\sum_{k=1}^{2N} \mathbb{1}_{[k \neq i]} \exp(s_{i,k}/\tau)}$
  $\mathcal{L} = \frac{1}{2N} \sum_{k=1}^N [\ell(2k-1, 2k) + \ell(2k, 2k-1)]$
  update networks f and g to minimize $\mathcal{L}$
end for
return encoder network $f(\cdot)$, and throw away $g(\cdot)$

FIGURE 18.13
Loss function for similarity maximization between vectors.

```
# Code for NT-Xent Loss function, explained in more detail in the article

tau = 0.05

def loss_function(a, b):
    a_norm = torch.norm(a, dim=1).reshape(-1, 1)
    a_cap = torch.div(a, a_norm)
    b_norm = torch.norm(b, dim=1).reshape(-1, 1)
    b_cap = torch.div(b, b_norm)
    a_cap_b_cap = torch.cat([a_cap, b_cap], dim=0)
    a_cap_b_cap_transpose = torch.t(a_cap_b_cap)
    b_cap_a_cap = torch.cat([b_cap, a_cap], dim=0)
    sim = torch.mm(a_cap_b_cap, a_cap_b_cap_transpose)
    sim_by_tau = torch.div(sim, tau)
    exp_sim_by_tau = torch.exp(sim_by_tau)
    sum_of_rows = torch.sum(exp_sim_by_tau, dim=1)
    exp_sim_by_tau_diag = torch.diag(exp_sim_by_tau)
    numerators = torch.exp(torch.div(torch.nn.CosineSimilarity()(a_cap_b_cap, b_cap_a_cap), tau))
    denominators = sum_of_rows - exp_sim_by_tau_diag
    num_by_den = torch.div(numerators, denominators)
    neglog_num_by_den = -torch.log(num_by_den)
    return torch.mean(neglog_num_by_den)
```

FIGURE 18.14
NT- Xent Loss function code.

```
# Boolean variable to control whether to train the linear classifier or not
LINEAR = True

class LinearNet(nn.Module):
    def __init__(self):
        super(LinearNet, self).__init__()
        self.fc1 = torch.nn.Linear(100, 5)

    def forward(self, x):
        x = self.fc1(x)
        return(x)
```

FIGURE 18.15
Linear classifier code.

train a batch size 256 using cloud GPU and carry out the decaying of learning rate with cosine decay schedule without restarts. We use SGD optimizer with square root learning rate scaling for analysis purpose as we have considered a small batch size and small number of epochs, but training with SGD becomes unstable with large batch size; in that case, LARS optimizer can be used [1]. Figure 18.14 shows the code for the loss function where tau is a temperature hyperparameter, which makes loss function more expressible and here, the similarity between two vectors a and b is the dot product of their respective unit vectors a_cap and b_cap.

18.3.3.6 *Training of ResNet-18 using NT- Xent Loss*

A linear classifier is trained on the representations obtained from the second last layer of projection head as the representation of this layer is better than that of the one after it, whose code is visible in Figure 18.15. The training and testing accuracy and losses versus number of epochs graph are generated which is discussed upon in the next section.

18.4 Results and Discussion

In this chapter, we have analyzed how different components act on our dataset to yield results, which helps us to draw a number of conclusions. Here, we will highlight and discuss the results and related inferences.

18.4.1 Data Augmentation: Original and Augmented Image

We went ahead and applied a combination of two data augmentation operations: first, random crop and resize and then color distortion as the second; two times for each image, to get two new images. Following is the result we obtained (Table 18.1).

TABLE 18.1

Two Images Each from "Donut" and "Bear" Category Obtained by Applying Composition of Random Crop and Resize and Color Distortion on the Original Picture

Original image	After applying composition of random crop and Resize with Color distortion one time	After applying composition of random crop and resize with Color distortion two times

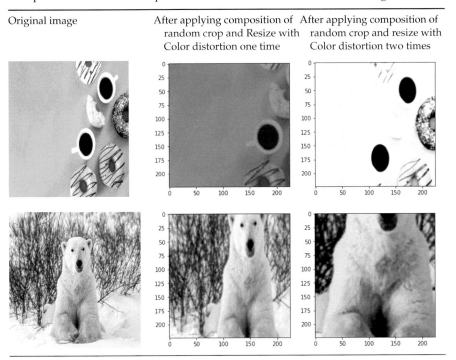

The conclusion here is that even though the positive pairs are getting identified efficiently, but there exists no single transformation obtained after data augmentation that is enough to learn good representations; but the above discussed composition is good for learning generalizable features.

18.4.2　Layers of Neural Network Base Encoder and Projection Head

After the augmentation of images, we fed each positive pair in a convolutional neural network having ResNet-18 architecture and two-layer projection head to project the representations further. Figure 18.11 showed how the last and ReLU activation layers were removed while the Figure 18.12 showed the removal of the last layer of the projection head. The t-SNE visualization of these above actions are shown in Tables 18.2 and 18.3.

> t-SNE visualization of training dataset t-SNE visualization of testing dataset

The conclusions drawn from the above discussion and results are as follows:

- Removal of hidden layer and ReLU activation before projection head is a better representation of the image vectors.
- Presence of clusters in above visualizations show that correct pairs of images are being identified.

TABLE 18.2

t-SNE Visualizations of the Last Layer Vectors of Train (10% of 1,250 =125) and Test (250) Images

t-SNE visualization of training dataset	t-SNE visualization of testing dataset

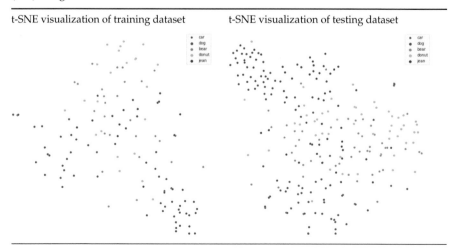

TABLE 18.3

t-SNE Visualizations of the Second Last Layer Vectors of Train (10% of 1250 =125) and Test (250) Images

t-SNE visualization of training dataset	t-SNE visualization of testing dataset

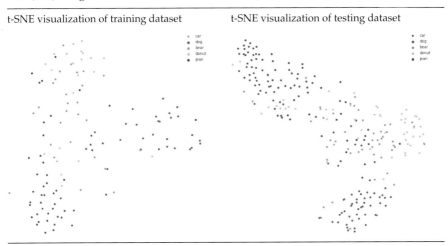

```
# Plot the training losses Graph and save it
fig = plt.figure(figsize=(10, 10))
sns.set_style('darkgrid')
plt.plot(losses_train)
plt.legend(['Training Losses'])
plt.savefig('losses.png')
plt.show()

# Store model and optimizer files
torch.save(resnet.state_dict(), 'results/model.pth')
torch.save(optimizer.state_dict(), 'results/optimizer.pth')
np.savez("results/lossesfile", np.array(losses_train))
```

FIGURE 18.16
Code to visualize training losses graph.

- Visualization of dataset on RHS of above tables are considered to be low rank [1] as they are poorly separate as compared to the dataset on the other side of the tables.

18.4.3 Training Losses

We modified the architecture of our Resnet-18 model by replacing top and last layers of it by fully connected layers and a classifier respectively. On training the Resnet-18 model, some training losses calculated by NT-Xent Loss function alongside the number of epochs used for training is visualized using matplotlib.pyplot (Figures 18.16 and 18.17).

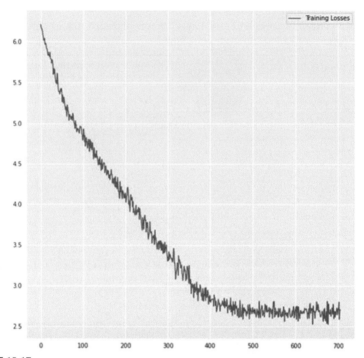

FIGURE 18.17
Training losses versus no. of epochs graph.

18.4.4 Training of ResNet-18 using NT-Xent Loss

A linear classifier is trained on the representations obtained from the second last layer of projection head as the representation of this layer is better than that of the one after it, whose code is visible in Figure 18.15. The training and testing accuracy and losses versus number of epochs (no. of epochs used here=20) graph are generated in Table 18.4.

On training of ResNet-18 using NT- Xent Loss, we obtain a test accuracy of 64.8 percent, which is 12 percent lower than that obtained from training ResNet-18 using supervised classifier which gives accuracy of 76.8 percent.

18.5 Conclusion and Future Work

In this work, we present a simple framework for contrastive visual representation learning, which is an improved technology we have used here for image recognition and classification. It is a high-performance model with high accuracy in which labeled samples are used for optimizing model parameters.

TABLE 18.4

Accuracy and Losses Graphs Plotted While Training a Linear Classifier on 10 Percent Labeled Training Data

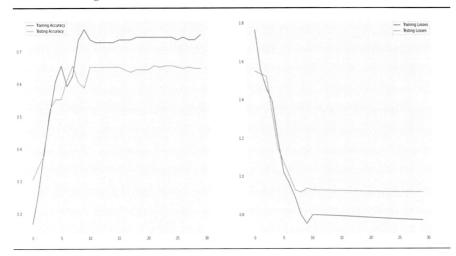

We take a close look at its components and demonstrate the results of various design choices. We build on previous approaches for self-supervised, semi-supervised, and transfer learning and integrating our results and analysis with it. Data augmentation, usage of a nonlinear head at the neural network's end, and a normalized loss function distinguish our method from normal supervised learning on ImageNet dataset.

With applications in facial recognition, driverless vehicles, medical disease detection, and even education, image recognition using SimCLR has enhanced online visibility. Thus, image recognition systems have a bright future.

Driverless vehicles, as well as amazing facial recognition mechanics, have already been introduced in a number of countries, allowing for quicker and more accurate object detection in real time. This was made possible by image recognition using the SimCLR algorithm. Hence, it has become easy to expect that the image recognition market will thrive in the upcoming years.

References

1. Ting Chen, Simon Kornblith, Mohammad Norouzi, and Geoffrey Hinton, "A simple framework for contrastive learning of visual representations", In the Proceedings of the 37th International Conference on Machine Learning, Vienna, Austria, July 2020.

2. Hadsell, R., Chopra, S., and LeCun, Y., "Dimensionality reduction by learning an invariant mapping", In 2006 IEEE Computer Society Conference on Computer Vision and Pattern Recognition (CVPR'06), volume 2, pp. 1735–1742. IEEE, 2006.

3. R Devon Hjelm, Alex Fedorov, Samuel Lavoie-Marchildon, Karan Grewal, Phil Bachman, Adam Trischler, and Yoshua Bengio, "Learning deep representations by mutual information estimation and maximization", *arXiv* preprint arXiv:1808.06670, August 2018.

4. Philip Bachman, R Devon Hjelm, and William Buchwalter, "Learning representations by maximizing mutual information across views", In the 33rd Conference on Neural Information Processing Systems (NeurIPS 2019), Vancouver, Canada, 2019.

5. Aaron van den Oord, Yazhe Li, and Oriol Vinyals, "Representation learning with contrastive predictive coding", arXiv preprint arXiv:1807.03748, 2018.

6. Olivier J. Henaff, Aravind Srinivas, Jeffrey De Fauw, Ali Razavi, Carl Doersch, S. M. Ali Eslami, and Aaron van den Oord, "Data-efficient image recognition with contrastive predictive coding", In the Proceedings of the 37th International Conference on Machine Learning, Vienna, Austria, 2020

7. Zhirong Wu, Yuanjun Xiong, Stella Yu, and Dahua Lin, "Unsupervised feature learning via non-parametric instance discrimination", In Proceedings of the IEEE Conference on Computer Vision and Pattern Recognition, pp. 3733–3742, 2018.

8. Kaiming He Haoqi Fan Yuxin Wu Saining Xie and Ross Girshick, "Momentum contrast for unsupervised visual representation learning", *arXiv* preprint arXiv:1911.05722, 2019.

9. Ishan Misra and Laurens van der Maaten, "Self-supervised learning of pretext-invariant representations", *arXiv* preprint arXiv:1912.01991, 2019.

10. Alexey Dosovitskiy, Philipp Fischer, Jost Tobias Springenberg, Martin Riedmiller, and Thomas Brox, "Discriminative unsupervised feature learning with exemplar convolutional neural networks", In Advances in Neural Information Processing Systems, pp. 766–774, 2014.

11. Yonglong Tian, Dilip Krishnan, and Phillip Isola, "Contrastive Multiview coding", *arXiv* preprint arXiv:1906.05849, 2019.

12. Ye, M., Zhang, X., Yuen, P. C., and Chang, S.-F. "Unsupervised embedding learning via invariant and spreading instance feature", In the Proceedings of the IEEE Conference on Computer Vision and Pattern Recognition, pp. 6210–6219, 2019.

13. Alex Krizhevsky, Ilya Sutskever, and Geoffrey E. Hinton, "Imagenet classification with deep convolutional neural networks", In Advances in Neural Information Processing Systems, pp. 1097–1105, 2012.

14. Christian Szegedy, Wei Liu, Yangqing Jia, Pierre Sermanet, Scott Reed, Dragomir Anguelov, Dumitru Erhan, Vincent Vanhoucke, and Andrew Rabinovich, "Going Deeper with Convolutions", In Proceedings of the IEEE Conference on Computer Vision and Pattern Recognition, pp. 1–9, 2015.

15. https://medium.com/analytics-vidhya/understanding-simclr-a-simple-framework-for-contrastive-learning-of-visual-representations-d544a9003f3c.

Index

Note: Page numbers in **bold** refer to tables and those in *italic* refer to figures.